Backyard Guide to
Insects & Spiders
of North America

NATIONAL
GEOGRAPHIC

Backyard Guide to
Insects & Spiders
of North America

ARTHUR V. EVANS

NATIONAL GEOGRAPHIC
WASHINGTON, D.C.

Contents

ABOUT THIS BOOK   6

BACKYARD BASICS

Insect Watching   10
Where & When   12
Attracting & Observing   16
Meet the Arthropods   18
Arthropod Life Cycles   24
What Arthropods Eat   30
Signs of Arthropod Life   36
Social Insects   42
Welcoming the Arthropods   46

GUIDE TO 150 SPECIES

Dragonflies & Damselflies   50
Cockroaches, Termites & Mantises   60
Earwigs   67
Grasshoppers, Katydids & Crickets   68
Stick Insects   80

True Bugs & Allies   83
Lacewings, Antlions & Allies   109
Beetles   112
Mosquitoes & Flies   140
Butterflies, Skippers & Moths   152
Wasps, Bees & Ants   183
Spiders, Ticks & Allies   203
Centipedes & Millipedes   219
Pillbugs & Sowbugs   221

APPENDIX

Glossary   224
Citizen Science
  & Other Resources   229
About the Author   231
Acknowledgments   231
Art Credits   232
Index   234

Native plants and their flowers provide food and shelter for butterflies, such as monarchs, and other insects and spiders. *previous pages:* A lady beetle explores a wildflower.

About This Book

In this book we introduce you to the 150 species of arthropods—insects, spiders, and kin—that you are most likely to find at home, inside your house or in your backyard. Most range far and wide throughout the United States and Canada. In this book we also introduce you to the ideas and vocabulary of entomology and the details of anatomy and behavior that scientists use to identify and study these fascinating creatures. Look around—look closer—and learn to enjoy these fascinating creatures with whom we share our world.

BACKYARD BASICS

These pages will orient you to the world of insects and spiders. They provide an overview of key facts about the places they live, the food they eat, and their stages of development. Basics of body parts and the scientific classification of insects, spiders, and kin by order also appear in this section. With this basic information on how to find and observe insects in your backyard, all you will need to add is a little curiosity.

GUIDE TO 150 SPECIES

Here you will get to know the 150 arthropods, organized by order (a scientific category) and each featured on a page of its own. They are grouped into these orders based on physical and other characteristics that support their shared evolutionary histories. Each species is represented by a photograph, a set of key facts, a description of its appearance and habits, a list of similar species (if appropriate), and an illustration highlighting some interesting detail of appearance or behavior. As an aid, vocabulary that may be unfamiliar is printed in **boldface** in the main text and defined in the Glossary, found on pages 224–228.

THEMATIC SIDEBARS

Occasionally there is a story to tell about insect or spider behavior that deserves its own page or pages. For these, we offer illustrated sidebars. Here you will find interesting facts, solid advice, and more details to look for as you observe the insects and spiders in your backyard.

GLOSSARY AND RESOURCES

Entomologists—the scientists who study insects—use a special vocabulary to describe the parts of the body and the behaviors of insects and spiders. We use this vocabulary throughout, believing that you learn even more about these creatures by learning how to describe them. Every term printed in **boldface** in the book is defined in the Glossary.

Here you will also find ideas and resources on how to participate in some of the many citizen science opportunities involved with studying insects and spiders. You can contribute to ongoing research as you observe and learn more each day.

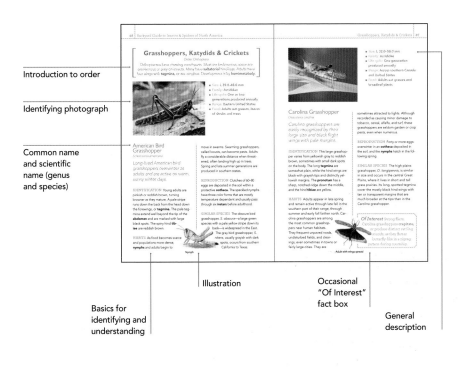

Introduction to order

Identifying photograph

Common name and scientific name (genus and species)

Basics for identifying and understanding

Illustration

Occasional "Of Interest" fact box

General description

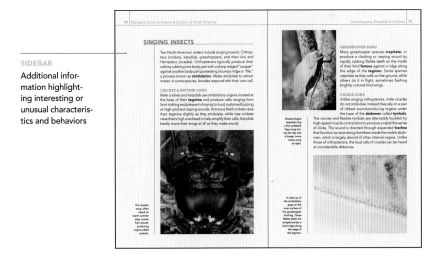

SIDEBAR
Additional information highlighting interesting or unusual characteristics and behaviors

BACKYARD BASICS

[Insect Watching]

*I*nsects, spiders, and other arthropods—the dominant animals on Earth—are at once strange and familiar. With skeletons worn on the outside of their bodies, six or more legs, and unblinking eyes, these animals appear quite alien to us. They are everywhere, yet we tend to pay them little notice.

GETTING TO KNOW ARTHROPODS

Our relationships with arthropods are long and complex. Some bite, sting, transmit disease-causing pathogens, attack our crops, or infest our food stores—and we spend billions of dollars annually to control them. Yet most species cause us little or no direct harm, and many are absolutely essential to a well-functioning environment, thus contributing directly to our very survival and well-being.

One of the best things about insects and spiders is that you don't have to go far to see them or need expensive equipment to enjoy them. To be sure, the greatest diversity of species is found in distant tropical rain forests, but there are plenty of amazing species worthy of attention that await discovery right in your backyard or a nearby garden or park.

EMBRACING DIVERSITY

Never before has there been a better time to embrace and celebrate arthropods and their diversity at home. Whether you live in the city, in suburbs, or out in the country, insects, spiders, and other arthropods are sure to be familiar animals. Even if you don't see them, evidence of their activities abounds.

Have you ever wondered about the identity of the animals that fill warm days and nights with the pleasant cacophony of buzzes, chirps, clicks, rasps, and whines? How do bees and wasps construct nests of wax, mud, and paper and do so with such precision? Why do spiders crisscross footpaths and decorate lawns and hedges with silk? If you have ever pondered these or any other aspects of the myriad multilegged creatures that live all around you, read on.

A colorful painted lady butterfly sips nectar from a thistle flower, a preferred food plant of its caterpillars.

[Where & When]

*U*nlike skittish birds and mammals, insects and spiders often allow you to get remarkably close. Keen eyes and an inquisitive mind are all that are needed to observe these animals as they mate, lay eggs, feed, and develop. Finding arthropods in your backyard doing all these things and more is easy if you know when and where to look.

TIME OF YEAR, TIME OF DAY

In North America, arthropod activity picks up in spring and is often timed with bud break and the appearance of flowers—as early as late February or March at lower elevations across the southern United States, Southwest, and southern California and as late as May or early June at higher elevations or in the northern United States and southern Canada.

Herbivorous insects appear with the sudden availability of plant food in the form of blossoms and soft, young leaves. Arthropod **predators** and **parasitoids** start emerging, too, to take advantage of these emerging herbivores.

The heat of summer marks the appearance of many singing insects: cicadas during the day and crickets and katydids at night. Arthropod activity drops off dramatically in late fall with

Large milkweed bug

Crab spider

the arrival of the first frosts. In winter arthropods are carefully tucked away in the soil or under bark to avoid freezing to death, although some species that are adapted to cooler temperatures are active during this time of year. Across the southern United States and in coastal southern California many insects and spiders remain active year-round.

Arthropod watching in your backyard and beyond is also dependent on time of day. Warm, sunny days are generally best for **diurnal** species, particularly after rains. Warm, humid nights, especially evenings with cloud cover or no moon, are conducive to **nocturnal** insect activity. **Crepuscular** species are briefly active just before dawn or dusk. Many of our large beetles and moths fly late in the evening, just before or after midnight.

Spring's warm weather and blooming plants make a perfect environment for herbivorous insects looking to feed on the new leaves.

WHERE TO LOOK

Eastern black swallowtail

Look for arthropods in well-watered habitats that are free of pesticides and have a variety of native plants. Tread lightly so these habitats continue to support insect and spider diversity.

FLOWERS AND VEGETATION
Flowers are incredibly attractive to nectar- and pollen-feeding insects as well as their **predators** and **parasites.** Carefully examine fruits, seedpods, cones, leaves, stems, branches, and trunks on various plants, day and night.

Immature greenhouse millipedes

FUNGI, MOSSES, AND LICHENS
Using a pocket magnifier loupe, carefully examine all surfaces of these organisms and the soil below for small arthropods, taking care not to damage them.

DECAYING SNAGS, LOGS, AND STUMPS
Moist, decaying wood attracts the greatest number of arthropod species, and their diversity continuously changes over time. Carefully pull back loose bark to check its underside and the exposed wood. Be sure to replace the bark whenever possible.

Water strider

BENEATH STONES AND DEBRIS
Arthropods frequently take shelter beneath stones, boards, and other objects and debris lying on the ground. Always return these objects back to their original position for the benefit of the organisms living there and the aesthetic appearance of the area.

Turn on the Lights
Nocturnal insects are attracted to porch lights, storefronts, and well-lit gas stations, especially in less developed areas. The bluish glow of mercury vapor streetlights is the most attractive. Some species will fly or crawl directly to the light, while others climb on nearby walls or plants. Larger beetles often remain in the nearby shadows. Insect activity at lights often attracts spiders and other predatory arthropods, too.

LEAF LITTER AND COMPOST

Accumulations of leaves and needles under trees and shrubs harbor many kinds of arthropods. Compost heaps and other piles of rotting vegetation are also used as shelters. Carefully raking through this material often reveals arthropods.

SHORELINES OF STREAMS, RIVERS, AND LAKES

Many arthropods take refuge under piles of debris that wash up along the shore, and many can be found among nearby vegetation. Floating debris on flowing and standing waters may harbor species swept from the shore by wind.

PONDS AND OTHER FRESHWATER WETLANDS

Search for aquatic and semiaquatic species in, on, and near ponds, bogs, swamps, and quiet pools along streams and rivers. Some species glide over the surface; others swim through the water column or cling to various submerged objects and substrates.

CARRION AND DUNG

Look under **feces** and dead animals in various stages of decay for specialized species unlikely to be found anywhere else. The arthropods found in these situations recycle nutrients, making them available to other organisms. Always exercise caution when examining animal remains and waste, and thoroughly wash your hands afterward.

Fallen trees, like this one in Pacific Rim National Park in western Canada, are home to many species of arthropods.

Attracting & Observing

Chances are there are plenty of insects and other arthropods already in your yard. By experimenting with different kinds of food as bait, you are likely to see many species new to you. With the right tools in hand, you can begin to observe and appreciate their lives close up.

SWEET AND NOT-SO-SWEET OFFERINGS

Insects are drawn to different kinds of baits, some sweet and others not. Overripe fruit attracts all kinds of insects—especially butterflies, flies, and wasps. Beer or wine, mixed with molasses and a pinch of brewer's yeast and set out in small, shallow dishes at the base of trees, often attract beetles and other insects drawn to sap. A concoction of overripe fruit, molasses, beer, and brown sugar painted on tree trunks at dusk on warm, humid days—a technique known as sugaring—may attract all kinds of **nocturnal** moths and beetles.

Canned pet food or fish, shrimp, and chicken parts set out to decompose will lure species that are typically attracted to carrion. Place these items under a board weighed down with bricks or cinder blocks, or wrap them

Binoculars: Get the best you can afford that allow a focus 8 feet away or less. Binoculars that are 8x42 or 10x42 are best for watching insects and spiders.

Pets for an hour: Children can enjoy insects up close by keeping them in a ventilated jar, but only for a short period of time.

A dish of ripe fruit is the perfect food for attracting monarchs and other butterflies.

up in hardware cloth with half-inch mesh so the rotting material has time to work its magic. In most areas, the waste of dogs and cats attracts very few kinds of insects.

Your success for attracting insects depends largely on local weather conditions. Experiment with different baits and methods of presentation to see what works best in your area.

Once you're attracting insects, you can sharpen your observation skills. Journaling is a good way to create a permanent record of your observations. Get in the habit of noting dates, climatic conditions, and attributes of the landscape such as prominent trees, shrubs, plants, and other natural features. Note the insect sounds you hear. A species' first and last appearances for the season are always notable. Over time, your journal will serve as a measure of your growing relationship with arthropods and deepening awareness of the natural world.

The sugar in overripe fruits will also draw insects like bees and wasps to your backyard.

[Meet the Arthropods]

*I*nsects, spiders, and their relatives belong to the animal phylum Arthropoda. Arthropods have tough outer skeletons, or **exoskeletons,** that are further divided into ringlike segments. Their jointed appendages—legs, mouthparts, and **antennae**—are also segmented, affording these structures greater flexibility. All four major living groups of arthropods, or **subphyla,** are represented in this guide.

INSECTS

SUBPHYLUM HEXAPODA, CLASS INSECTA

Adult insects—including ants, bees, beetles, butterflies, cicadas, dragonflies, flies, moths, and wasps—have three distinct body regions: head, **thorax,** and **abdomen.**

The head bears the primary sensory organs and mouthparts. They have one pair of antennae, a pair of **compound eyes** with multiple lenses, and sometimes one or more simple eyes, or **ocelli.** The mouthparts are usually adapted for chewing, piercing-sucking, or lapping.

The thorax has three pairs of legs and often one or two pairs of wings. An intricate network of veins usually supports the membranous wings. The forewings of some insects are modified and uniformly thick with veins **(tegmina),** half thick and half membranous with veins **(hemelytra),** or leathery or hardened without any trace veins **(elytra).** Insects are further classified into orders, 11 of which appear in this book.

Odonata: twelve-spotted skimmer

Dictyoptera: American cockroach

Orthoptera: eastern lubber grasshopper

COMMON INSECT ORDERS

ORDER	COMMON NAME	DEVELOPMENT	MOUTHPARTS	WINGS
Odonata	Dragonflies, Damselflies	hemimetaboly	chewing	4
Dictyoptera	Mantises, Cockroaches, Termites	hemimetaboly	chewing	4, 0 tegmina
Dermaptera	Earwigs	hemimetaboly	chewing	4, 0 tegmina
Orthoptera	Grasshoppers, Crickets, Katydids	hemimetaboly	chewing	4, 0 tegmina
Phasmida	Stick insects	hemimetaboly	chewing	0, 4 tegmina
Hemiptera	True bugs, Cicadas, Aphids, Hoppers, kin	hemimetaboly	piercing-sucking	4, 0 hemelytra
Neuroptera	Antlions, Lacewings, kin	holometaboly	chewing	4
Coleoptera	Beetles	holometaboly	chewing	4, 2 elytra
Diptera	Flies, Mosquitoes, Midges	holometaboly	sponging, piercing-sucking	2
Lepidoptera	Butterflies, Skippers, Moths	holometaboly	siphoning	4, 0
Hymenoptera	Ants, Bees, Wasps	holometaboly	chewing	4, 0

Hemiptera: green stink bug

Neuroptera: green lacewing

Coleoptera: Japanese beetle

Diptera: common green bottle fly

Lepidoptera: western tiger swallowtail

Hymenoptera: paper wasp

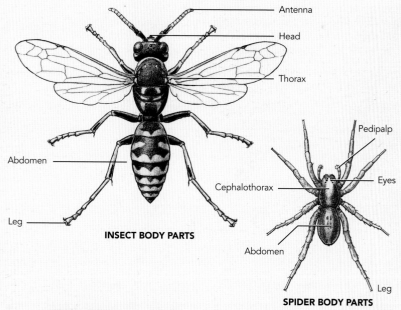

INSECT BODY PARTS

SPIDER BODY PARTS

SPIDERS & KIN

A close-up view shows four of the eight eyes, the hairy pedipalps, and the iridescent turquoise chelicerae of the daring jumping spider.

SUBPHYLUM CHELICERATA, CLASS ARACHNIDA
Spiders and their kin, including mites and ticks, are classified as arachnids. They have two major body regions; in spiders, the two body regions—**cephalothorax** and **abdomen**—are quite distinctive. Arachnids lack **antennae,** but some species use their front legs primarily as sensory organs. The pincher-like mouthparts, or **chelicerae,** are modified into fangs in spiders. Adults have eight legs. Three orders of arachnids are covered in this book.

SPIDERS & KIN				
ORDER	COMMON NAME	CHELICERAE	ABDOMEN	LEGS
Araneae	Spiders, Tarantulas	fangs	usually unsegmented	variable
Acarina	Mites, Ticks	pincher-like (mites), piercing (ticks)	unsegmented	short
Opiliones	Harvestmen	pincher-like	segmented	long

CENTIPEDES & MILLIPEDES

SUBPHYLUM MYRIAPODA, CLASSES CHILOPODA & DIPLOPODA
Centipedes, millipedes, and other myriapods have one pair of antennae, simple or compound eyes, chewing mouthparts, and a long segmented body trunk supported by numerous pairs of legs. They are usually found in moist terrestrial habitats.

Centipedes (Chilopoda) are somewhat flattened and run quickly when threatened. Their relatively long legs extend out to the sides, one pair for each of the leg-bearing body segments. Depending on the species, centipedes have between 15 and 193 leg-bearing segments. The fanglike front legs are held under the head like mouthparts, helping centipedes to capture and subdue small invertebrates by injecting venom. The last pair of legs extends backward. They are used as sensory organs, not for walking.

Though they may not be welcome guests in your home, house centipedes do feed on household pests such as bedbugs, termites, silverfish, and cockroaches.

Millipedes (Diplopoda) are cylindrical or somewhat flattened and move slowly compared with centipedes. They have short legs underneath the body, with two pairs on most body segments. Each body segment, or **diplosegment,** is a fusion of two segments. Depending on the species, millipedes have 11 to 192 leg-bearing segments. When threatened, most curl up into a flat coil, although some can roll up into a ball like a pillbug. Some diplosegments bear **repugnatorial glands** on the sides that secrete volatile and noxious defensive fluids that will stain skin. Millipedes are slow-moving **detritivores** that spend their lives burrowing through soil and litter, ingesting plant debris and converting it into humus.

ROLY POLYS

SUBPHYLUM CRUSTACEA, CLASS MALACOSTRACA

The staggering diversity of crustaceans surpasses that of insects, complicating attempts to summarize their body forms succinctly. Most species are marine or aquatic. Malocostracan crustaceans include lobsters, crabs, crayfish, shrimp—many of which are considered delicacies—and isopods, an order that includes pillbugs and sowbugs, collectively known as roly polys.

A pillbug in defensive posture

Common rough woodlice

[Arthropod Life Cycles]

*There are two basic types of development in arthropods: **direct development**, in which the hatching offspring resemble the adults, and **indirect development**, in which the young, or larvae, look and behave very differently from the adults.*

IN THE BEGINNING

In general, insects, spiders, and other arthropods begin their lives as fertilized eggs, although a few species reproduce **asexually** by **parthenogenesis.** In ants, bees, and wasps, unfertilized eggs always develop into males, while fertilized eggs become females. Aphids (p. 106) alternate between generations of winged males and females, which reproduce sexually, and wingless females, which produce more wingless females asexually that are genetically identical to themselves.

In a honey bee brood, the queen lays an egg into each cell. In a healthy brood, the queen can lay between 1,000 and 2,000 eggs per day.

Mexican bean beetle larva

Colorado potato beetle larva

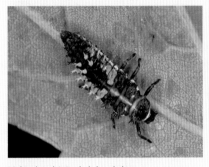

Multicolored Asian lady beetle larva

Green lacewing larva

Most arthropods are capable of reproduction only as adults. Females sometimes deposit their eggs randomly or, more often, on or near a suitable food source for their young. A few species retain their eggs inside their bodies, where they hatch and develop, eventually emerging as larvae.

Marvels of Molting Young arthropods quickly outgrow their stiff **exoskeletons** and shed them in a series of molts that are hormonally controlled. In insects, the stage between each **molt** is called an **instar.** Each successive instar is usually similar to the last in form, but larger. After shedding the old exoskeleton, the arthropod has a new soft, pale one that undergoes a chemical hardening and darkening process akin to the tanning of leather over the next few hours or days.

DEVELOPING THROUGH STAGES

Direct development occurs in several insect orders and in other arthropods, including arachnids, millipedes, and centipedes.

Upon hatching from their eggs, spiderlings resemble the adults but lack functional, fully segmented appendages. They cannot move or feed, and they remain inside the egg sac. After the next molt, they leave the egg sac and begin to hunt for food. Depending on the species, juvenile spiders molt three to nine times before reaching adulthood.

Some arthropods undergo **anamorphosis** and grow additional legs and/or body segments as they develop. Ticks (pp. 215, 218) and mites hatch from their eggs as six-legged larvae and become eight-legged nymphs at the first molt. Millipedes (p. 220) hatch with six legs and add additional legs and body segments with subsequent molts up to adulthood.

Primitive wingless insects such as silverfish and bristletails undergo **ametaboly** (no metamorphosis). The young hatch from the egg looking very much like a small version of the adult, but they are incapable of reproduction. Unlike most other adult insects that stop molting when they reach adulthood, these insects continue to molt for the rest of their lives.

Over the course of its life, the Mexican bean beetle matures through several distinctive forms, shown here clockwise from upper right. Starting as eggs, they develop into spiny yellow larvae. Mature larvae attach themselves to leaves and transform into pupae. Adults emerge in about a week.

A dragonfly naiad

Emerging from the naiad shell

A mature blue corporal dragonfly

Winged insects, such as dragonflies, mantises, cock-roaches, grasshoppers, and true bugs, develop by **hemimetaboly,** or gradual metamorphosis. When they hatch from eggs, they resemble adults but lack wings. The immature stages of dragonflies and damselflies, mayflies, and stoneflies are all aquatic; they are called **naiads.** In all the other hemimetabolous orders of insects (whether the immature stages are aquatic or terrestrial), the young are called **nymphs.**

As naiads and nymphs reach maturity, their developing wings become apparent on the outside of the body, if they are present in the adult.

Modes of Metamorphosis There are three basic types of metamorphosis: ametaboly, with little or no change from one stage to another; hemimetaboly, in which immature individuals—nymphs or naiads, for instance—transform directly into mature ones; and holometaboly, or complete metamorphosis, in which the life cycle goes from egg to larva to pupa to adult. Larval and adult holometabolous insects often lead very different lives and seldom compete with one another.

Most insects, including beetles, flies, butterflies and moths, ants, bees, and wasps, undergo a type of indirect development known as **holometaboly** (complete metamorphosis). Holometabolous species undergo four distinct stages of development: egg, larva, pupa, and adult. The larvae, variously known as **caterpillars,** grubs, or **maggots,** look and behave very differently from the adults. Wing development during the larval stage occurs internally and becomes evident externally, along with other adult structures, only in the pupa.

Caterpillars of the black swallowtail butterfly at different stages of development

Some holometabolous insects are **parasitoids**—that is, they are **parasites** that ultimately kill their hosts—and they undergo a special form of development called **hypermetamorphosis,** in which the larval instars appear in two or more very different forms.

For example, in blister beetles (p. 132), the first instar is a leggy, active larva called a **triungulin** that is adapted for seeking the proper host. Once a host is located, the triungulin molts into a decidedly sedentary and short-legged grub, followed by a molt into a fat, legless grub. It eventually develops into a more active, short-legged larva and spends most of its time preparing for pupation. Some hypermetamorphic flies and wasps have legless first-instar larvae, called planidia.

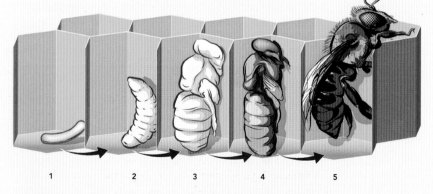

1   2   3   4   5

Queen bees lay their eggs (1) in individual cells in the brood. The egg matures into a larva (2), which is fed by worker bees in the hive. Over several days, the larva develops into a pupa (3), which begins to take on a more bee-like appearance (4). When the bee is fully mature (5), it emerges from the cell to join the hive.

When mature, a monarch caterpillar attaches itself to a branch (1) and molts into a chrysalis (2). The adult emerges 9 to 14 days later and, once the wings are fully developed (7), is ready to take flight.

LIFE CYCLES

The length of time it takes for an arthropod life to complete its life cycle, from egg to adult, varies depending on species and conditions.

Some species, such as mosquitoes (p. 141), may require barely more than a week to go from egg to adult and may produce multiple generations within a single year. Others typically produce just one or two generations annually, overwintering in the life stage (egg, nymph, larva, pupa, or adult) best adapted to cold temperatures. At the other end of the spectrum, periodical cicadas (p. 97) may take up to 13 or 17 years to reach adulthood. And, while most spiders live for about a year, some female tarantulas are known to live 20 to 30 years.

Within a single species that is widely distributed in North America, southern populations may produce three or more generations annually, while northern populations may produce only one generation every one, two, or more years.

[What Arthropods Eat]

Insects, spiders, and other arthropods consume an amazing variety of foods. Some are specialists, eating only plants or animals, while others eat both. Some consume decaying organic tissues and animal waste. Within each category, some specialize in liquid or solid foods, as evidenced by their modified mouthparts.

HERBIVORES

Herbivorous arthropods consume all parts of a plant: flowers, fruits, seeds, buds, leaves, stems, branches, trunks, and roots. Many flower-visiting insects sip nectar, eat pollen, or collect pollen for their young. The pollination services rendered by these species are essential to human agriculture.

On the other hand, leaf-chewing insects may cause considerable damage. They selectively nibble away soft tissues, leaving only a skeleton-like network of veins behind, or they hungrily chomp around leaf edges, leaving irregular notches behind. Some may cause severe defoliation.

Oak leaf galls house wasp larvae.

Getting to Know Galls

Some insects and mites create **galls**—swellings or deformities of plant growth that can at first glance look like a bud, fruit, or nut. Certain insect species initiate gall formation chemically as they lay eggs or feed on the plant. They zero in on plants in just one genus or species—broadleaf trees and goldenrod are favorites in North America. They even target a specific location—bud, leaf, stem, or other plant structure. As the gall grows, it provides the insect or mite developing inside with food and shelter, but it seldom damages the host plant.

Insects with chewing mouthparts also attack flowers, fruits, and seeds. The most conspicuous of these in North America include grasshoppers, beetles, and **caterpillars,** as well as the larvae of wasps known as sawflies.

Less conspicuously, some insect larvae feed within plant tissues, chewing tunnels just below the leaf's surface, a behavior known as mining that leaves blotches or meandering tunnels behind. Wood-borers, especially beetle larvae, selectively consume the inner bark, **phloem, sapwood,** or **heartwood** of trees. Some insects feed and develop within the safety of specialized plant structures known as **galls.**

Bees and butterflies have modified mouthparts that enable them to sip nectar from flowers, while the piercing-sucking mouthparts of plant-feeding Hemiptera (true bugs) permit them to draw sap directly from a plant's vascular tissues.

European honey bee with pollen

Silver-spotted skipper sipping nectar

Chokecherry midge maggots in fruit

Grasshopper eating wheat

Large milkweed bug nymphs sucking plant sap

Aspen leafminers tunneling through leaf

CARNIVORES

Arthropod carnivores include **predators, parasitoids,** and **parasites.** Predators hunt, trap or ambush and kill their prey, usually other arthropods. Wheel and assassin bugs (pp. 86–87), fiery searchers (p. 112), and house centipedes (p. 219) all actively search for insect prey. Mantises (p. 66) and crab spiders (p. 211) typically sit and wait, motionless, until a prey item comes within reach. The aquatic **naiads** of dragonflies and damselflies (pp. 50–59) ambush prey on the bottoms of pools and ponds and dart after flying insect prey, consuming them on the wing. Conspicuous, fast flying, and keen-eyed dragonflies perch atop vegetation.

Some insects and spiders build traps to ambush prey. The larvae of some antlions (p. 110), known as doodlebugs, spiral backward down into the sand, using their wedge-shaped bodies and flat heads like plows and shovels to construct conical pit traps with slippery slides that catch ants and other small crawling insects.

Many spiders construct intricate webs to entangle prey. Orbweavers (p. 206) build webs with dry strands of silk radiating from the center and a continuous spiral of silk coated with sticky microscopic droplets to capture flying insects. The seemingly chaotic cobwebs spun by widow spiders actually contain different structural features depending on the spider's needs. Hungry widows (p. 205) build webs with sticky lines of silk to ensnare ground-dwelling arthropods, while well-fed spiders surround themselves in a protective tangle of nonsticky silk threads.

A caterpillar hunter beetle eats a gypsy moth caterpillar.

Tiny parasitic wasp cocoons attach to a tomato hornworm.

A golden orb-web spider wraps its prey with silk.

A southern black widow with a cricket in its web

Parasitoid larvae usually kill their host. Parasites also depend on a single host but rarely kill or significantly harm them. Parasites that live on the outside of the host's body are called **ectoparasites,** while those developing inside the host are **endoparasites.** The larvae of the tiger bee fly (p. 144) are **ectoparasitoids** that ultimately kill their host, the grubs of eastern carpenter bees (p. 190). The **endoparasitoid** larvae of wasps in the genus *Cotesia* develop inside and in tandem with that of tomato hornworms (p. 176). The invaded host stays alive long enough to meet the wasp larvae's nutritional needs. Finally the larvae chew their way out of the hornworm's body, ready to spin cocoons and pupate.

A Chinese mantid devours a grasshopper.

DETRITIVORES

Arthropods that consume carrion, leaf litter, decaying flesh, and dung—the **detritivores**—are nature's recyclers. Along with fungus and bacteria, they help break down these materials into progressively smaller particles, ultimately returning nutrients to the soil and thus making them available for use by living organisms.

For example, dung beetles that remove and bury dung for their young play a big role in keeping our environment clean by limiting breeding sites of pestiferous flies. Cockroaches (pp. 60–62) and roly polys (pp. 221–222) also break down dead plant and animal materials into smaller particles via their **feces,** which are then attacked and broken down further by fungi and bacteria. Wood-boring beetles, termites, carpenter ants, and other insects that feed on dead wood help to convert stumps, snags, and fallen logs into fertile soil. Carrion beetles, blow flies, and flesh flies (p. 150) are saprophages that specialize in feeding on decaying flesh.

A dung beetle gathers up a ball of feces, to be used for food or a brooding chamber.

Maggots, Murder, and Other Mayhem
Forensic entomology is the study of insect biology as it relates to legal proceedings. Forensic entomologists carefully study blow flies and other carrion-feeding insects associated with a corpse at a crime scene to determine the time of death and whether or not the body has been moved. Insect-related evidence is also used to link a suspect with the victim. Forensic entomologists also collect saprophagic insects and other arthropod-related evidence to investigate cases of child abuse, elder neglect, automobile accidents, and plane crashes.

Decaying leaves and other organic matter on the forest floor are important sources of food and shelter for many insects and spiders.

Signs of Arthropod Life

As your fascination with insects and spiders grows, so will your awareness of the evidence of their feeding, reproductive, and nesting activities, some of which are more evident after leaves begin to drop in autumn.

FINDING EGGS

Many insects hide their eggs singly in the soil, leaf litter, or rotten wood, while others deposit them in plain sight on leaves and twigs. The greater angle-wing katydid (p. 72) attaches eggs in rows to leaf and twig edges. Asian tiger mosquitoes (p. 141) lay dark, rice-like eggs a few at a time along the edges of small bodies of water. Wheel bugs and assassin bugs (pp. 86–87) deposit hexagonal batches of bottle-shaped eggs on twigs and leaves. Egg masses of eastern tent **caterpillars** look like a dark, varnished coating that completely surrounds the twig.

Cockroaches and mantises (pp. 60–66) lay eggs in protective cases called **oothecae.** The oothecae of cockroaches are hard, purse-like cases that vary in color from tan to dark brown, contain up to four dozen eggs, dropped near adequate supplies of food and water. Depending on the species, mantis oothecae may be long and slender or squat and round, attached to branches and walls, and contain as many as 200 eggs. The **nymphs** emerge simultaneously from a row of narrow openings, each borne on a single strand of silk.

The angle-wing katydid attaches rows of flat, oval eggs to leaves and twigs.

Cockroaches leave hard, purse-like oothecae, or egg cases, near food and water.

The eastern tent caterpillar wraps its dark, varnished egg case around host twigs.

SPIDER EGG SACS

Spider eggs sacs are variously shaped. A single egg sac may contain more than 2,000 eggs. Some spiders, like the long-bodied cellar spider, carry their egg sacs in their **chelicerae** until they hatch.

Wolf spiders carry their spherical egg sacs at the tip of the **abdomen** while they hunt for food. Hatching spiders climb up on the female spider's back. Female green lynx spiders (p. 210) attach their egg cases to vegetation and will defend it until the spiderlings hatch. The somewhat spherical egg sacs of the common house spider, southern black widow, and black and yellow garden spider (pp. 204–205, 208) are enveloped by parchmentlike silk and suspended in the web.

Black and yellow garden spider egg sac

Female (L) and male (R) black widows with egg sac

Green lynx spider with egg sac

Cellar spider with egg sac

COCOONS

Often you may find a cocoon, within which a moth is developing. Learning what each species' cocoon looks like can help you know which species will eventually emerge.

Eastern tent **caterpillars** (p. 168) pupate within an elongate and tapered cocoon covered in yellowish or whitish powder and surrounded by webbing. Giant silk moths produce cocoons made of very tough parchmentlike silk that may incorporate leaves. Cocoons of cecropia moths (p. 175) reach 8 to 10 centimeters in length, may or may not include leaves, and are usually attached to a twig along one side. Polyphemus moth (p. 173) cocoons are typically oval, wrapped in one or more leaves, and sometimes found attached to a twig by a narrow strap of silk.

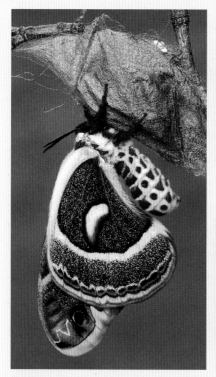

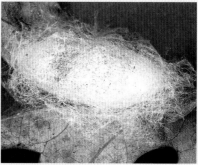

Eastern tent caterpillar cocoon

A cecropia moth recently emerged from its cocoon   Polyphemus moth cocoon

INSECT NESTS

Ants, bees, and wasps make various structures in which to raise their brood. Some dig burrows, while others secrete wax

or gather mud for use as construction materials. The nests of bees and wasps may contain multiple cells provisioned with food for the young. Some of these nests are quite distinctive and are useful for identifying the builder.

Anthill entrance

ANTS

Anthills—the aboveground evidence of an ants' nest—often appear at the edges of large rocks and sidewalks or along cracks in pavement. Depending on the species, they often appear as craterlike mounds of loosely packed, fine, uniformly sized grains of soil neatly arranged around a single nest entrance, or as irregular piles with multiple entrances.

Worker bees on brood cells

BEES

Native bumble bees (p. 187) mix pollen with wax to construct cup-shaped cells on the bottom of their subterranean nests to store food and serve as brood chambers. Introduced European honey bees (p. 186) build vertical two-layered combs, mostly of pure wax, to store honey and raise brood. Feral honey bees establish their nests in tree hollows, under protected overhangs, inside walls, or even out in the open, on tree branches.

But not all bees build nests with wax. Solitary eastern carpenter bees (p. 190) chew perfectly round entrance holes about 1 cm wide in the sound wood of railings, decks, and eaves.

Paper wasps building a nest

WASPS

Mud daubers (pp. 183, 185) build their nests with balls of mud collected from nearby puddles. Inside these nests are multiple chambers, each with paralyzed spiders provisioned as food for a single wasp larva. Potter wasps (p. 193) stock their juglike nests with **caterpillars** and other larvae as food for their young.

Social wasps build nests using a paperlike material called **carton** that they make by chewing wood into pulp and mixing it with **labial gland** secretions. Paper wasps (p. 194) build a horizontal disk that consists of a layer of honeycomb-like cells suspended by a single stalk. Yellowjackets and hornets build several layers of horizontal cells and cover them with multiple layers of carton. Hornets (p. 195) typically build their nests aboveground, while yellowjackets (p. 196) construct their nests in cavities in the ground, in hollow logs, or inside other low, concealed sites. The cells in all of these carton nests open downward and are used solely to brood larvae.

SPIDERWEBS

Spiders are unique among all arthropods in that they produce and use silk throughout their lives. Some spiders produce as many as seven different kinds of silk, each for a different use. They use silk as safety and climbing lines, in the construction of egg sacs and webs, for building temporary bivouacs or lining burrows, and for wrapping and storing prey.

The most familiar use of silk by spiders is for capturing their prey through web construction. Depending on the species, spiderwebs consist of snares, nets, funnels, sheets, or other designs to trap crawling or flying arthropods. The owner of the web is alerted that its web has ensnared a potential meal by vibrations transmitted through the silk threads as the prey struggles to escape.

Many spiders also produce a silken dragline as they walk or jump, to avoid losing their footing and falling to the ground. They regularly tack the dragline to the substrate with more silk, like a mountaineer securing safety lines by hammering pitons into the rock.

Orb weavers, such as this black and yellow garden spider (p. 208), build elaborate webs with dry strands of silk radiating from the center, bound with a continuous spiral of silk coated with sticky microscopic droplets that capture flying insect prey.

Newly hatched spiders may release long, slender threads into the wind as parachutes upon which to carry themselves, a means of dispersal called **ballooning.** Most are transported only a few yards, although some ballooning spiders have made it hundreds of miles out to sea. Ballooning spiders have little control over the direction and distance that they are transported, although those with longer lines of silk tend to travel farther. Mass ballooning involving hundreds or thousands of individual spiderlings is triggered by developmental and environmental cues.

[Social Insects]

The vast majority of insects and spiders lead solitary lives, coming together briefly only to mate, after which the females lay eggs and depart long before their progeny hatch. *Eusocial* (truly social) behavior occurs only in the insect orders Dictyoptera and Hymenoptera: All termites and ants, and some bees and wasps are eusocial, with duties divvied up among different forms, or castes, including reproductive females and males and sterile workers.

QUEENS & KINGS

Virgin **queen** and king termites and ants (those that are reproductive) are winged; all but male ants shed their wings after mating.

A colony is founded by at least one mated queen, who is the mother of the entire colony. Some ants, such as red imported fire ants, live in colonies with multiple queens.

A queen may live for only one year (bumble bees, yellowjackets, paper wasps, hornets), for up to five years (honey bees), or even longer (termites, ants). A long-lived queen has the potential to lay millions of eggs over her lifetime. Depending on the species, the queen may be much larger than the workers or not all that different in appearance.

Reproducing males, or kings, contribute little to the daily functioning of the colony other than fertilizing the queen. In termites, they remain with the queen, but among ants, bees, and wasps, they die soon after mating.

The winged reproductives of eastern subterranean termites are called alates.

Worker eastern subterranean termites are rarely seen without breaking open a piece of infested wood.

WORKERS

The vast majority of individuals within a colony are winged (bees, wasps) or wingless (termites, ants) workers. Multiple generations work together, caring for the queen, raising their siblings, gathering food, and repairing, expanding, and defending the nest.

Younger workers of ants, bees, and wasps tend to function as nursemaids and housekeepers, while older workers leave the nest to forage for food or defend the nest. In some ants and termites, larger workers have large heads with powerful **mandibles** adapted for defense: They are known as soldiers.

The workers of ants, bees, and wasps are all adult females, while those of termites include male and females, both adults and **nymphs.**

Nuptial Swarms The first rains of spring and summer trigger mating flights of ants and termites. Future king and **queen** termites mate and shed their wings as they begin their lives together. Mated queen ants also shed their wings, but the males simply die.

THE NATURE OF COLONIES

Colonies of bumble bees, hornets, and yellowjackets are founded by a lone queen. Paper wasp nests may be founded by multiple queens but are eventually dominated by one. All these queens, along with their colonies' workers and **drones,** die at the end of a season, but the new generation of mated queens will persist through the winter and start their own colonies the following spring. European honey bee, ant, and termite colonies will persist for several years, as long as their queens continue to lay eggs.

European hornet workers fly about their nest entrance.

The complex tasks performed by the members of the colony are achieved in the absence of any central control exerted by the queen or other individuals in the colony. These examples of swarm intelligence are of great interest to computer scientists studying the principles of self-organization and how apply them to various human endeavors, including Internet search systems, factory production and product distribution, telecommunication networks, and other complex economic activities.

COMMUNICATION

A significant part of the success story of social insects is their ability to communicate with one another employing chemicals known as **pheromones.**

Ants greet one another by gently tapping with their **antennae** to detect the specific chemical odor of their fellow nest mates. Through grooming and licking, the worker ants ingest pheromones produced by the queen, behavior that keeps the colony together and functioning smoothly. These pheromones also accelerate or inhibit worker activities, including suppressing their own limited ability to reproduce.

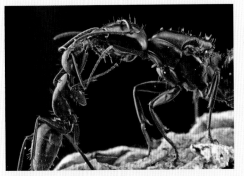

A black carpenter ant worker feeds the queen.

Worker ants disseminate the queens' pheromones throughout the colony orally, while termites ingest the anal secretions of one another, activities called **trophallaxis.** Workers produce and use pheromones for a number of purposes: to recruit fellow workers to defend the colony, to coordinate nest building and repair, and to convey information on the whereabouts of food and water.

HONEY BEE DANCES

Honey bees communicate without pheromones through their complex dances. Scout bees returning from foraging engage in figure eight or waggle dances that recruit nest mates to

Successful foraging honey bee workers return to the hive to recruit others to find food and water.

visit a productive patch of flowers. Recent research suggests, though, that the dance simply triggers foraging behavior rather than describing the path to the site. Honey bees accumulate electric charges as they fly; those charges are then emitted as electric fields as they dance. It has been suggested that these fields stimulate organs within the bees' antennae and thus play a role in their communication.

The Advantages of Being Eusocial Although they make up only 2 percent of the world's known insect species, **eusocial** insects occupy up to half the world's insect **biomass.** They play powerful roles in nature via pollination (bees), seed dispersal (ants), soil aeration (ants), wood decomposition (termites), and predation of insects and other arthropods (ants, wasps).

following pages: Certain garden flowers, such as this echinacea, will attract more butterflies.

[Welcoming the Arthropods]

*T*o many, the only good arthropod is a dead one. Entire aisles at gardening shops and hardware stores are dedicated to selling poisons specifically formulated to kill them. Sadly, using these products tends to encourage pest species by killing less resistant **predators** and **parasitoids** first. Better to find natural deterrents and at the same time marshal your curiosity about the lives of these creatures.

RESTORING NATURE

Matters are complicated further by urban and suburban gardens composed of exotic species planted solely for aesthetic purposes that are food deserts for native arthropods and other wildlife. In these diminished habitats, native arthropods and animals have less to eat, fewer places to live, and a dearth of places to hide, reproduce, or engage in other life processes critical to their survival. It's time to think of the plants in our yards not as ornaments but as active parts of the web of life that supply food and shelter for local wildlife

How can you make a difference at home and in your community? Go native! Yards, gardens, schoolyards, parks, and road medians filled with carefully selected native plant species appropriate for the region not only encourage arthropod diversity but also attract birds and other wildlife that depend on insects, spiders, and their relatives. Give nature a hand by planting native flowering perennials, including shrubs and trees.

A rich diversity of arthropods, especially **caterpillars,** provides food for nesting birds and other wildlife as well. Almost anything you do to enhance your yard, school, or park, whether it involves planting natives or simply reducing mowing and the use of pesticides, will benefit native wildlife and enhance your sense of accomplishment and well-being.

The words of entomologist and author Douglas Tallamy sum it up best: "Garden as if life depends on it!"

GUIDE
TO 150 SPECIES

Dragonflies & Damselflies

Order Odonata

At rest, dragonflies hold their wings out, while most damselflies fold them over their backs. Adults and aquatic **naiads** have chewing mouthparts. Development is by **hemimetaboly.**

- Size: L 68.0–84.0 mm
- Family: **Aeshnidae**
- Life cycle: **One generation produced annually**
- Range: **Across southern Canada and United States**
- Food: **Adults prey on small insects**

Common Green Darner

Anax junius

The common green darner is among the largest and most familiar dragonflies in North America.

IDENTIFICATION The head is bluish above with a bull's-eye pattern in front of the eyes. The male's large, thick **thorax** is plain green without any side stripes, while the **abdomen** is blue and green with a dark line along the top. Females and immature males look similar but have a reddish-brown abdomen.

HABITS Males patrol the edges of ponds, lakes, brackish waters, and slow streams in spring and summer, searching for females. Both sexes engage in small feeding swarms throughout the day, preying mostly on small insects. Common green darners are one of the few migratory dragonflies in North America, flying south in the fall.

REPRODUCTION Mating pairs of only this darner species remain together during **oviposition.** They hang among low grasses and weeds along the shore. Females alight on floating mats of vegetation to lay their eggs, usually with a male in tandem.

SIMILAR SPECIES Males of the southwestern giant darners (*A. walsinghami*) have a longer, stripeless blue abdomen that arches in flight. In the East, comet darners (*A. longipes*) resemble the female common green darner, but their red abdomens lack a top stripe.

Naiad

- Size: L 51.0–58.0 mm
- Family: **Libellulidae**
- Life cycle: **One generation produced annually**
- Range: **Across southern Canada and United States**
- Food: **Adults prey on small insects**

Twelve-Spotted Skimmer
Libellula pulchella

Large and conspicuous, twelve-spotted skimmers are mesmerizing to watch as they patrol along the shores of ponds and engage in aerial acrobatics with rival males.

IDENTIFICATION Both male and female twelve-spotted skimmers have three large black spots on each of four wings: hence the name. Mature males have white wing patches between the black spots, pale stripes on the sides of the **thorax,** and a somewhat chalky gray **abdomen.** Females lack white wing patches and have yellow **thoracic** stripes. The abdomens of mature females and immature males are brown with yellow stripes along the sides.

HABITS On sunny days, twelve-spotted skimmers fly almost continuously in search of prey and mates. They breed in sunny bogs, marshes, lakes,

slow-moving streams, and other permanent bodies of standing water with abundant emergent vegetation. Individuals in search of prey are sometimes found in yards and fields. Males are aggressive toward other males and will fly vertical loops around them. They occasionally rest in prominent spots near the shore, especially on the very top of tall, bare branches. In the Northeast, this species sometimes engages in migratory swarms, flying southward along the coast in fall.

REPRODUCTION Copulation occurs briefly on the wing. Occasionally guarded by their mate, females energetically tap their abdomens on the surface, sometimes splashing water well in front of them, as they deposit eggs near aquatic vegetation.

SIMILAR SPECIES The western eight-spotted skimmer (*L. forensis*) looks similar but lacks dark wing tips. Female twelve-spotted skimmers resemble those of the common whitetail (*Plathemis lydia*) but lack the white zigzag stripes on the sides of the abdomen.

Naiad

- Size: L 50.0–55.0 mm
- Family: **Libellulidae**
- Life cycle: **One generation produced annually**
- Range: **Across southern Canada and United States**
- Food: **Adults prey on small insects**

Black Saddlebags
Tramea lacerata

Black saddlebags are so named because of the distinctive dark patches on their hind wings.

IDENTIFICATION These large, slender dragonflies are mostly black with unmarked forewings and a broad black patch at the base of each hind wing. The bodies of immature adults are lighter in color with purplish or brown hues and five pairs of pale spots on top of the **abdomen** that usually darken with age. The last pair of spots may persist to maturity, especially on females.

HABITS Powerful fliers, black saddlebags are found mostly at lower elevations around ponds, lakes, irrigation ditches, and vegetated sloughs from spring through early fall. Males fly low and fast, patrolling their territories with their abdomens held straight or bent slightly downward. Not only are male black saddlebags aggressive toward males

of their own species, but also they tend to be aggressive toward males of other dragonfly species. Nonbreeding individuals are found in a wide variety of open habitats, including suburban yards.

REPRODUCTION While flying in tandem, *Tramea* males briefly release their female partners so they can drop to the surface of the water, tap the water with the tips of their abdomens, and thereby release a batch of eggs. The male then quickly recaptures the female, and each pair will repeat this unique egg-laying behavior several times before finally parting ways.

SIMILAR SPECIES In western North America, red saddlebag (*T. onusta*) males have a red face, a red abdomen, and red wing veins that make the broad wing patches appear red. The females are similar overall but duller. In eastern North America, male Carolina saddlebags (*T. carolina*) appear similarly red, but they have a purplish face; females are duller and have black wing patches.

Naiad

- Size: **L 47.0–50.0 mm**
- Family: **Libellulidae**
- Life cycle: **One generation produced annually**
- Range: **Across southern Canada and United States**
- Food: **Adults prey on small insects**

Wandering Glider
Pantala flavescens

Wandering gliders spend their days patrolling over open habitats, including schoolyards and parking lots in summer.

IDENTIFICATION Adults appear uniformly yellowish to yellow-orange in flight, and they have broad, unmarked wings. The sexes look similar, but the females lack the male's reddish eyes and orange face.

HABITS Known as the globe skimmer in Europe, this species breeds on both sides of the Atlantic and Indian Oceans. Swarms of these insects are reported flying out over the sea, where they must keep flying for several days and nights before reaching distant continents and isolated oceanic islands. Their broad wings enable them to ride currents of air high above the water that are associated with monsoon fronts and other weather systems.

In North America, large concentrations of wandering gliders occur mostly in the East, but they have also been reported in the Southwest. They are attracted to temporary rain pools and drainage ditches, and they will sometimes dart through swarms of small insects, eating them on the wing.

REPRODUCTION Mated pairs fly in tandem in long straight lines while the females deposit eggs by tapping the surfaces of temporary and seemingly inappropriate bodies of water, such as rain puddles and temporary wetlands, even swimming pools and garden ponds. They will even lay their eggs on the hot, shiny surfaces of parked cars! **Naiads** develop rapidly in mostly non-vegetated and fishless bodies of water. Northern populations migrate southward in fall, even flying over the ocean.

SIMILAR SPECIES Spotwinged gliders (*P. hymenea*) are similar in size, but they are darker in color and have a distinct spot at the base of each of their four wings.

Naiad

- Size: **L 42.0–48.0 mm**
- Family: **Libellulidae**
- Life cycle: **One generation produced annually**
- Range: **Across southern Canada and United States**
- Food: **Adults prey on small mosquitoes, flies, and other small insects**

Common Whitetail
Plathemis lydia

Common whitetails are widespread in North America. Colors and wing patterns distinguish the sexes.

IDENTIFICATION Males and females have short and chunky bodies and a pair of oblique yellow stripes on the side of the **thorax,** but their wing patterns differ. Males' wings have a broad black band near the tip and a short black streak at the base. When mature, the male's **abdomen** is white, sometimes appearing bluish. Females' wings each have three dark spots, including one at the tip. Their abdomen, similar to that of an immature male, is brown with yellow or white diagonal spots down the sides.

HABITS Common whitetails are usually found away from water in spring and summer, resting on the ground, rocks, and logs, or perching on low branches. Mature males establish territories along the edges of muddy marshes, ponds, lakes, and slow-moving

rivers, displaying their thick white abdomens as a warning to rival males. Over the water, they elevate their abdomens while facing off with other males.

REPRODUCTION Attended by their mates, females lay their eggs while tapping their abdomens on the water near floating or emergent vegetation. As many as 50 eggs are dropped at a time with each flick. Covered in algae, the sprawling dark brown **naiads** live on the bottom where they ambush small aquatic invertebrates as well as small tadpoles and fish. They are tolerant of varying water quality.

SIMILAR SPECIES Male southwestern whitetails (*P. subornata*) have splashes of white on their inner wings, while the females lack black wing tips.

Of Interest Male whitetails are very territorial and will warn off rival males intruding on their borders by holding their white abdomens in the air.

Adult female

HOW DRAGONFLIES DO IT

The mating behavior of Odonata is unique among insects and fascinating to observe.

TERRITORIAL

Many male dragonflies guard territories that vary in size according to species and population density. Territories generally contain good egg-laying sites that will attract one or more females. Male dragonflies defend their territory with aggressive behaviors intended to ward off rival males as well as those of other species. Some engage in visual displays that involve flashing their bright body colors and patterned wings. Male damselflies generally do not establish territories, with ebony jewelwings and their relatives being the exception. Females entering these territories are alerting their owners of their readiness to mate.

COURTSHIP ON THE WING

Before coupling, the male transfers sperm from the tip of his **abdomen** to a secondary sexual organ at the base of the abdo-

men. With special clasping organs at the tip of his abdomen, he grabs the back of the female's head (dragonflies) or **pronotum** (damselflies). The pair is now "in tandem." The male then curves his abdomen downward, allowing the female to curl her abdomen forward to bring her sexual organs at the tip into contact with his secondary sexual organs. The heart-shaped outline with their coupled bodies is called the wheel-position.

A pair of damselflies in the wheel-position—a distinctive mating behavior shared by members of Odonata

PROTECTING PATERNITY

Upon inserting this organ into the female's genital opening, males of many species first remove sperm deposited by the previous male before supplanting it with their own, thus assuring their paternity of her fertilized eggs. To prevent their sperm from suffering a similar fate, these males remain with their mates by standing guard nearby, hovering overhead, or remaining in tandem as she lays her eggs.

- Size: L 30.0–37.5 mm
- Family: Coenagrionidae
- Life cycle: One generation produced annually
- Range: Western North America
- Food: Adults prey on small insects

Vivid Dancer
Argia vivida

Vivid dancers are seemingly ubiquitous damselflies, most often encountered along streams running through interior valleys or flowing down foothill canyons.

IDENTIFICATION Although named for the mature male's brilliant blue color, immature males are mostly whitish, becoming lavender gray and then tan before they reach maturity. Age and temperature will change the colors of mature males from deep violet blue to dull slate blue. The male's **thorax** has black stripes; the **abdomen** has black rings that increase in width toward the rear. Abdominal segments 8–10 are bright blue. Depending on age and temperature, females are similarly variable in color as the males, but their black abdominal rings are narrower.

HABITS Vivid dancers are most active from spring through late summer, but these damselflies are sometimes encountered year-round in some areas. Males bask in the morning on streamside rocks and vegetation. Although they sometimes occur around larger bodies of water, such as ponds, lakes, and rivers, this species is more likely to breed in the shallow, rocky, and vegetated streams that feed or drain these larger bodies of water.

This large blue dancer species has adapted to disturbed habitats and is sometimes found breeding in artificial drains in suburban neighborhoods, as well as in irrigation ditches in rural areas.

REPRODUCTION After mating, females in tandem with males lay eggs just below the water surface on the stems of submerged vegetation.

SIMILAR SPECIES Along the Pacific coast states, mature males resemble Emma's dancers (*A. emma*), Aztec dancers (*A. nahuana*), and California dancers (*A. argioides*), but vivid dancers are generally bluer and more robust than these other species.

Adult female

- Size: **L 25.0–30.0 mm**
- Family: **Coenagrionidae**
- Life cycle: **One generation produced annually**
- Range: **Western North America**
- Food: **Adults prey on small insects**

Pacific Forktail

Ischnura cervula

Pacific forktails are the first among damselflies or dragonflies to appear in spring and are often encountered year-round in California.

IDENTIFICATION Males are mostly black above, except for the eyespots, four spots on top of the **thorax,** and abdominal segments 8 and 9, which are all pale blue. Their green eyes are capped in black, while the sides of the thorax are pale blue. Females are similar in color but with striped thoraxes. Young females range from pale tan to lavender gray, while older individuals become darker and **pruinose,** or lightly coated with a dusting of whitish or grayish wax.

HABITS Pacific forktails live in a wide variety of habitats but are most abundant at lower elevations along creeks, small streams, springs, roadside ditches, marshes, ponds, lakes, and other wetlands covered with algae and duckweed. It is not unusual to see these damselflies wandering well away from water to forage for insects in backyards and weedy fields.

REPRODUCTION Males perch among low, dense vegetation and will their **abdomens** up in the air at the approach of rival males. Females, often unguarded by males, **oviposit** in the stems of sedges and other aquatic plants.

SIMILAR SPECIES Older female Pacific forktails are difficult to distinguish from other species of *Ischnura* that also become dark and pruinose with age or cold temperatures. Males resemble both black-fronted (*I. denticollis*) and San Francisco (*I. gemina*) forktails but are distinguished from these and all other damselflies by the four spots on top of the thorax.

Of Interest Male Pacific forktails claim clumps of aquatic vegetation as mating territories and display their blue-tipped abdomens at one another.

Adult female

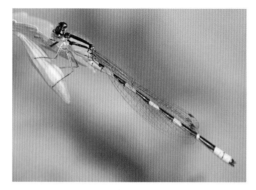

- Size: **L 25.0–40.0 mm**
- Family: **Coenagrionidae**
- Life cycle: **One generation produced annually**
- Range: **Across southern Canada and United States**
- Food: **Adults eat small insects**

Familiar Bluet
Enallagma civile

Well-known and widespread, the familiar bluet is a common sight along the edges of swamps, ponds, and other standing water habitats

IDENTIFICATION The striking males are blue with blue teardrop-shaped eyespots, a black-striped **thorax,** and a mostly blue **abdomen** with black rings. Abdominal segments 6 and 7 are mostly black, while 8 and 9 are blue. Immature females are brownish or tan, while mature individuals are bluish or greenish. The top of the abdomen is black in all color forms.

HABITS Familiar bluets take wing beginning in spring and remain active through late fall. They are abundant around standing water, especially ponds of all sorts, irrigation ditches, reservoirs, and rice fields but are less evident along small, slow-moving streams and in temporary wetlands. Individuals are often found flying over lawns in gardens and parks, well away from water.

REPRODUCTION After mating, tandem pairs seek emergent stems and rushes just above the water surface. The female crawls down these stems and into the water, remaining submerged for up to an hour as she lays her eggs, while the male stands guard above, high and dry.

SIMILAR SPECIES Male familiar bluets are among the bluest of this large species. In the East, they are nearly identical to the coastal Atlantic bluet (*E. doubledayi*), the males of which sometimes have their blue eyespots connected to form a bar across the head. Female bluets resemble several western *Enallagma*, especially the arroyo (*E. praevarum*) and tule (*E. carunculatum*) bluets.

Of Interest Unlike other species of bluets in the genus *Enallagma*, the widespread familiar bluet is most active in summer and remains on the wing well into autumn.

Adult female

- Size: **L 37.0–57.0 mm**
- Family: **Calopterygidae**
- Life cycle: **One generation produced annually**
- Range: **Eastern North America**
- Food: **Adults prey on small insects**

Ebony Jewelwing
Calopteryx maculata

The fluttering, dipping flight of ebony jewelwings is almost butterfly-like as they move along shaded streams in late spring and summer.

IDENTIFICATION Their slender bodies are deep metallic blue-green with dark brown wings that appear black, each tipped with a single white spot (female) or not (male). The long, slender legs are black and spiny.

HABITS Ebony jewelwings spend most of their days flying along shady, sandy-bottomed woodland streams and rivers but are sometimes found considerable distances away from water. Small groups of males and females engage in displays, flitting back and forth among the low, sun-dappled streamside vegetation before perching and slowly opening and closing their wings.

REPRODUCTION Males claim and defend territories around floating or submerged patches of vegetation, sometimes for a week at a time, greeting females with a courtship display. Males without territories will attempt to mate with females in the territories of other males. After several minutes of copulation, the females lay their eggs in small roots and submerged vegetation, sometime immersing themselves completely in the water, as the male stands guard nearby. The slender-bodied **naiads** live among submerged vegetation, roots, and debris in riffle zones. The average adult life span is about two or three weeks.

SIMILAR SPECIES Five species of *Calopteryx* occur in Canada and the United States, most in the East. All have metallic bodies, but only *C. maculata* has completely dark wings.

Naiad

Of Interest Ebony jewelwings are among the most intensely studied odonates in North America and have been the subject of numerous behavioral and ecological studies.

Cockroaches, Termites & Mantises

Order Dictyoptera

Dictyopterans are terrestrial insects with chewing mouthparts and four wings, if wings are present. Mantises and most cockroaches lead solitary lives, but termites are social. Development is by **hemimetaboly.**

- Size: L 34.0–53.0 mm
- Family: **Blattidae**
- Life cycle: **One generation produced every two years**
- Range: **North America**
- Food: **Adults eat plant and animal tissues**

American Cockroach

Periplaneta americana

Native to tropical Africa, American cockroaches, or palmetto bugs, are nearly cosmopolitan.

IDENTIFICATION They are shiny reddish brown overall, with a paler **pronotum** marked with a dark blotch slightly notched in front. The brown **tegmina** cover the **abdomen,** extending well beyond it in the male. **Nymphs** are mostly reddish brown, with a somewhat striped appearance.

HABITS Adults and nymphs scavenge food and foul stored products, and leave a strong odor where they live. They are pests in homes, heated basements of larger buildings, and food preparation facilities such as restaurants, grocery stores, hotels, and hospitals, where they seek high humidity areas; also common in sewers. Adults are weak fliers and only reluctantly fly.

REPRODUCTION Females produce up to 90 **oothecae** in a lifetime, each containing 14–16 eggs. The life cycle, from egg to adult, takes about 600 days, and adults may live up to another 400 days.

SIMILAR SPECIES Brown cockroaches (*P. brunnea*), less common, have an indistinctly marked pronotum. The southeastern smokybrown cockroach (*P. fuliginosa*) has a black pronotum; the Australian cockroach (*P. australsiae*) of Florida and southern Texas has a yellow pronotum with two black blotches and yellow stripes on the leading edges of the tegmina.

Ootheca (egg case)

- Size: L 25.0–32.0 mm
- Family: Blattidae
- Life cycle: One generation produced in two or more years
- Range: North America
- Food: Adults eat plant and animal tissues

Oriental Cockroach

Blatta orientalis

Originally from North Africa, oriental cockroaches are significant year-round house pests.

IDENTIFICATION These shiny black cockroaches, also known as water bugs, have fully developed **tegmina** that (in males) cover only three-fourths of the **abdomen** or (in females) are short, veined wing pads. Newly hatched **nymphs** are dark reddish brown and become blackish as they age.

HABITS Oriental cockroaches frequently enter homes and apartments through sewer pipes and are sometimes trapped in sinks, bathtubs, and shower basins. They prefer to occupy secluded places that are warm and humid, such as garages, basements, crawl spaces behind radiators and under floors, and water meter vaults buried in alleys and along parkways. Dense infestations of these insects are sometimes first detected by their pungent "roachy" odor. In warmer climates, they are commonly found living outdoors in hollow stumps and tree holes, hiding in leaf litter or under stones, or walking on sidewalks on warm, humid evenings.

SIMILAR SPECIES Turkestan cockroaches *(B. lateralis)* first appeared at a military installation in California during the 1970s, where they were likely introduced accidentally via military equipment returning from the Middle East. Males are uniformly pale brown with tegmina with cream-colored margins that extend past the abdomen. The mostly dark brown females have only wing pads, each with a pale stripe along the leading edge. This species lives primarily outdoors and is found around homes from central and southern California to western Texas; it is very abundant in the southern half of Arizona. Both Oriental and Turkestan cockroaches are sometimes found together in southern California.

Adult female

- Size: **L 11.0–16.0 mm**
- Family: **Ectobiidae**
- Life cycle: **Several generations produced annually**
- Range: **North America**
- Food: **Adults eat plant and animal tissues**

German Cockroach
Blatella germanica

A native of southeastern Asia, German cockroaches infest households in North America and throughout the world.

IDENTIFICATION These tan or pale brown insects have a distinct pair of dark stripes on the **pronotum.** Adult males are paler than females, and the tips of their **abdomens** are tapered, rather than broad. The **nymphs** also have striped pronota. Young nymphs are dark with a pale spot on their backs, while older nymphs have a short, broad pale stripe down the **thorax** and base of the abdomen.

HABITS German cockroaches are intolerant of cold and prefer to live indoors, where they seek places with optimal heat and moisture. They often infest kitchens and are frequently found during the day hiding in walls and other narrow spaces, especially around refrigerators, stoves, and sinks. Bathrooms sometimes become infested, too, if they share a common wall with the kitchen. Finding these normally **nocturnal** insects during the day may indicate a severe infestation.

REPRODUCTION A single female may produce four to five **oothecae,** each containing up to 40 or more eggs. She carries it on the tip of her abdomen for about a month and drops it off days before the eggs hatch. The nymphs take two to six months to reach adulthood.

SIMILAR SPECIES Asian cockroaches *(B. asahinai)* strongly resemble German cockroaches, but they prefer to live outdoors in shady mulched or composted areas. Originally from Japan, this species was first reported in Florida in 1986 and is now widespread in the southern Atlantic and Gulf Coast states.

Of Interest The beetle cockroach, *Diploptera punctata*, gives live birth. Scientists are studying the "milk" it produces to nourish its young as a possible superfood for humans.

Nymph

WHAT GOOD ARE COCKROACHES?

Of the world's 4,000 or more species of cockroaches, perhaps only 10 percent are considered pests. Yet we have allowed the habits of this small minority to tarnish an entire group of fascinating insects.

COCKROACH APPRECIATION

The vast majority of species live in tropical forests, woodlands, deserts, and many other habitats where they are a significant part of the F-B-I (fungus-bacteria-insects) that play a role in the recycling of nutrients. These agents of decomposition help break down plant and animal materials, making them easier to utilize as food by all sorts of organisms. A few species are even essential for the pollination of some tropical plants. Cockroaches also serve as research animals: Not only do they help us answer questions about insect behavior, anatomy, and physiology, but also their bodies continually inspire the development of new technologies.

Many cockroach species play an important role in the environment or are of great scientific interest.

SANITATION ENGINEERS & FOOD BANKS

As they feed, cockroaches join forces with other invertebrate scavengers to help recycle the remains of dead organisms and convert it into protein within their bodies. Protein-rich cockroaches are important sources of nutrition for birds, reptiles, amphibians, and fish, as well as for bats, monkeys, and other mammals that prey on them. Fungi and bacteria reduce cockroach waste into its most basic components and return them to the soil, where they will be taken up by other living organisms.

RESEARCH

Cheap, easy to keep, and prolific breeders in captivity, cockroaches make ideal experimental animals with seemingly endless applications. For example, cockroach studies have increased our general understanding of nervous system growth and development. Engineers study how cockroaches move to design and build equally nimble robots for rescue missions on Earth and for exploratory missions to Mars and beyond.

- Size: L reproductives 23.0–26.0 mm, including wings; soldiers 15.0–20.0 mm; false workers 13.0–20.0 mm
- Family: Archotermopsidae
- Life cycle: Colonies take several years to produce alates
- Range: Coastal British Columbia to California
- Food: Adults eat decaying wood of conifers

Pacific Dampwood Termite
Zootermopsis angusticollis

Pacific dampwood termites are North America's largest termite. Mature colonies of these social insects may number up to 4,000, including false workers, soldiers, and **reproductives.**

IDENTIFICATION The light yellowish brown and winged reproductives (kings, **queens**), or **alates,** are found in colonies year-round. False workers **(pseudergates)** are pale-yellowish male and female **nymphs** that may develop into either soldiers or reproductives. Yellowish- to reddish-brown soldiers use their well-developed **mandibles** to defend the colony from ants and other predatory arthropods.

HABITS Mating swarms are commonly attracted to lights, sometimes in large numbers. Colonies reside in galleries chewed in conifer stumps and logs, especially in moist coastal habitats. The walls of their galleries have a rough, almost velvety texture. They also infest timbers used to make homes, outbuildings, utility poles, and fence posts, and pilings along the Pacific Coast. Wet wood in contact with the ground is particularly susceptible to attack.

REPRODUCTION Mating swarms occur on warm afternoons and evenings, especially after rains, in the late summer. Their long and dark wings extend behind them, measuring twice the length of their bodies, and the wings detach after mating.

SIMILAR SPECIES The genus *Zootermopsis* contains only three species, all found in western North America. The smaller and darker Nevada dampwood termite, *Z. nevadensis*, shares the same distribution as *Z. angusticollis* but is much more common at higher elevations, where habitats are cooler and drier. *Zootermopsis laticeps* occurs in the mountain forests of central and southeastern Arizona and southwestern New Mexico.

Soldier

■ Size: L reproductives 10.0 mm, including wings; soldiers 5.0–7.0 mm
■ Family: Rhinotermitidae
■ Life cycle: Colonies take five or more years to produce reproductives
■ Range: Southern Ontario and eastern United States to Colorado
■ Food: Adults eat decaying wood

Eastern Subterranean Termite
Reticulitermes flavipes

The most common and important structural pest in eastern North America, eastern subterranean termites infest any type of wood or wood product.

IDENTIFICATION The pale workers are **nymphs;** they develop into soldiers with large, rectangular heads bearing a well-developed pair of **mandibles** that lack teeth on their internal margins. False workers, or **pseudergates,** resemble workers and may eventually develop into black, **alate** kings and **queens, or reproductives.** Their four long, grayish-brown wings are supported by a network of fine hairless veins.

HABITS Alates swarm outdoors in spring to mate but may emerge during winter inside heated buildings and die, wings attached, in sunlit windowsills, bathtubs, and sinks. Mated termites shed their wings just before establishing a new colony together. Colonies reach maturity in 5–10 years and may contain from 20,000 to 5 million individuals. They build their nests in the ground, where they build an extensive system of foraging tubes made of soil and bits of wood. These tubes, extending up tree trunks and across foundations, maintain the connection between the termite's food supply and the subterranean nest. They follow the wood's grain, eating the softer springwood and ignoring the harder summerwood behind, thus the layered look of infested timbers.

SIMILAR SPECIES Seven species of *Reticulitermes* occur in the United States, five of which are found in the Southeast. The western subterranean termite, *R. hesperus,* occurs throughout the West.

Of Interest Winged termites in the home indicate an infestation. Their four wings are of equal length, unlike those of ants. They have broad waists, while those of ants are narrow and wasplike.

Winged reproductive

- Size: L 83.0–104.0 mm
- Family: Mantidae
- Life cycle: One generation produced annually
- Range: Southern Ontario and Québec, throughout eastern United States and in California
- Food: Adults eat insects

Chinese Mantis
Tenodera sinensis

Introduced into the United States from China in 1896, the Chinese mantis is the largest of its kind in North America.

IDENTIFICATION These large insects are mostly green or brown, or green with brown margins on the **tegmina.** The underside of the **prothorax** is yellow between the **raptorial** forelegs. The tegmina extend beyond the length of the slender male's **abdomen** but don't quite reach the tip in the heavy-bodied female. At night, the Chinese mantis' **compound eyes** appear dark.

HABITS Adult females are especially common near flowers in gardens, parks, old fields, and roadsides in late summer and early fall, where they capture and consume large numbers of insects to assure proper egg development. They will lay up to 200 eggs within a foamy **ootheca** attached to a branch or other surface; the foamy medium soon hardens into a protective case. Gardeners and farmers purchase the spherical oothecae and put them in gardens and fields to take advantage of the voracious appetites of hatching **nymphs** in spring.

SIMILAR SPECIES Another introduced species, the narrow-winged mantis, *T. angustipennis,* is similar in appearance to the Chinese mantis, but it has a bright-orange spot between its raptorial forelegs. The green or gray Carolina mantis, *Stagmomantis carolina,* is native to the eastern United States; the green California mantis, *S. californica,* is widespread in the West. Heavy-bodied *Stagmomantis* females have short wings that cover only two-thirds of the abdomen.

Of Interest Mantises are opportunistic **predators** that eat anything they can capture. On occasion, individuals have been known to stake out bird feeders in order to catch and consume hummingbirds.

Ootheca (egg case)

Earwigs

Order Dermaptera

Earwigs have antlike heads, short **tegmina,** and **abdomens** tipped with pincherlike **cerci.** Omnivorous adults and **nymphs** have chewing mouthparts. Adults have four wings or are wingless. Development is by **hemimetaboly.**

- Size: L 9.0—17.0 mm
- Family: Forficulidae
- Life cycle: One or two generations produced annually
- Range: Across southern Canada and throughout United States
- Food: Adults eat plant tissues and small insects

European Earwig

Forficula auricularia

European earwigs are native to Europe, western Asia, and northern Africa. Their earliest report in North America was from Seattle, Washington, in 1926.

IDENTIFICATION They are flattened, reddish to blackish brown, with pale legs and **tegmina,** the tips of which are even paler. Although fully winged, they seldom fly. Each **antenna** consists of 14 **antennomeres.** Males are smaller with curved forcepslike **cerci,** while the larger females have cerci that are relatively straight.

HABITS Typically **nocturnal,** adults and **nymphs** hide in cool, dark, narrow spaces in flowers, buried in leaf litter, or under loose bark during the day. When insect prey becomes scarce, European earwigs shift to a diet of flowers and plant materials, sometimes becoming serious pests in greenhouses. Adults overwinter underground in well-drained soils or in aggregations under bark, hollow stems, and other frost-free places.

REPRODUCTION In spring, females dig chambers to lay and brood eggs and remain with hatched young for several days.

SIMILAR SPECIES The larger riparian or striped earwig *(Labidura riparia),* common across the southern United States, has two stripes on the **pronotum** and 25–30 antennomeres. Commonly attracted to lights at night in desert areas, it produces a disagreeable odor when disturbed.

Adult female

Grasshoppers, Katydids & Crickets

Order Orthoptera

Orthopterans have chewing mouthparts. Most are herbivorous; some are omnivorous or prey on insects. Many have **saltatorial** hind legs. Adults have four wings with **tegmina,** or are wingless. Development is by **hemimetaboly.**

- Size: L 39.0–48.0 mm
- Family: **Acrididae**
- Life cycle: **One or two generations produced annually**
- Range: **Eastern United States**
- Food: **Adults eat grasses, leaves of shrubs and trees**

American Bird Grasshopper

Schistocerca americana

Long-lived American bird grasshoppers overwinter as adults and are active on warm, sunny winter days.

IDENTIFICATION Young adults are pinkish or reddish brown, turning browner as they mature. A pale stripe runs down the back from the head down the forewings, or **tegmina.** The pale tegmina extend well beyond the tip of the **abdomen** and are marked with large black spots. The spiny hind **tibiae** are reddish brown.

HABITS As food becomes scarce and populations more dense, **nymphs** and adults begin to move in swarms. Swarming grasshoppers, called locusts, can become pests. Adults fly a considerable distance when threatened, often landing high up in trees. Spring and late summer generations are produced in southern states.

REPRODUCTION Clutches of 60–80 eggs are deposited in the soil within a protective **ootheca.** The speckled nymphs have three color forms that are mostly temperature dependent and usually pass through six **instars** before adulthood.

SIMILAR SPECIES The obscure bird grasshopper, *S. obscura*—a large green species with a pale yellow stripe down its back—is widespread in the East. The gray bird grasshopper, *S. nitens,* usually grayish with dark spots, occurs from southern California to Texas.

Nymph

- Size: L 32.0–58.0 mm
- Family: **Acrididae**
- Life cycle: **One generation produced annually**
- Range: **Across southern Canada and United States**
- Food: **Adults eat grasses and broadleaf plants**

Carolina Grasshopper
Dissosteira carolina

Carolina grasshoppers are easily recognized by their large size and black flight wings with pale margins.

IDENTIFICATION This large grasshopper varies from yellowish gray to reddish brown, sometimes with small dark spots on the body. The long **tegmina** are somewhat plain, while the hind wings are black with grayish tips and distinctly yellowish margins. The **pronotum** has a sharp, notched ridge down the middle, and the hind **tibiae** are yellow.

HABITS Adults appear in late spring and remain active through late fall in the southern part of their range, through summer and early fall farther north. Carolina grasshoppers are among the most common grasshoppers near human habitats. They frequent unpaved roads, undisturbed fields, and clearings, even sometimes in towns or fairly large cities. They are sometimes attracted to lights. Although recorded as causing minor damage to tobacco, cereal, alfalfa, and turf, these grasshoppers are seldom garden or crop pests, even when numerous.

REPRODUCTION Forty or more eggs overwinter in an **ootheca** deposited in the soil, and the **nymphs** hatch in the following spring.

SIMILAR SPECIES The high plains grasshopper, *D. longipennis,* is similar in size and occurs in the central Great Plains, where it lives in short and tall grass prairies. Its long, spotted tegmina cover the mostly black hind wings with tan or transparent margins that are much broader at the tips than in the Carolina grasshopper.

Of Interest Strong fliers, Carolina grasshoppers **crepitate,** or produce distinct rattling sounds, as they flutter butterfly-like in a zigzag pattern during courtship.

Adult with wings spread

- Size: L 28.0–50.0 mm
- Family: Acrididae
- Life cycle: One generation produced annually
- Range: Southern Ontario, most of United States
- Food: Adults eat leaves of many kinds of plants

Differential Grasshopper
Melanoplus differentialis

The differential grasshopper is common in moist vegetated habitats in natural areas, weedy vacant lots in urban settings, and cultivated croplands.

IDENTIFICATION Greenish or brownish with black markings, differential grasshoppers become darker with age, with black grooves visible on the **pronotum** and unmarked **tegmina.** The **saltatorial** hind legs have inverted black chevrons on the **femora** and yellowish **tibiae.**

HABITS Differential grasshoppers eat many kinds of plants, including grasses, **forbs,** shrubs, and trees, but thrive best on broadleaf plants and especially corn, cotton, and fruit trees. Adults and **nymphs** feed during the day and climb high up in vegetation to rest at night. They often move several miles in destructive swarms that ruin gardens, crops,

vineyards, and orchards. Differential grasshoppers are most common between the Rocky Mountains and the Mississippi River, with only scattered populations to the east and west. They are absent in the Pacific Northwest, extreme Northeast, and southeastern Coastal Plain.

REPRODUCTION Males are generally smaller than females. Females deposit up to six fragile and curved **ootheca** in the soil, each containing up to 200 eggs. Overwintered eggs hatch in early summer, and the nymphs reach adulthood in a month, feeding mostly on grains and alfalfa.

SIMILAR SPECIES The genus *Melanoplus* includes more than 200 species in America north of Mexico, many of which are important garden and agricultural pests. The recently extinct Rocky Mountain locust, *M. spretus,* was once widespread but is now known only from preserved specimens, including those found in a Montana glacier.

Female laying eggs

- Size: **L 43.0–80.0 mm**
- Family: **Acrididae**
- Life cycle: **One generation produced annually**
- Range: **North Carolina to Florida, west to Texas**
- Food: **Adults eat leaves of low-growing broadleaf plants**

Eastern Lubber Grasshopper

Romalea microptera

One of the best-known insects in the Southeast, the flightless eastern lubber grasshopper has several colorful monikers: Georgia thumper, devil's horse, and graveyard grasshopper.

IDENTIFICATION The robust, short-winged adults are black with yellow markings or yellow with black markings. They have slender legs and are poor jumpers.

HABITS Adults appear in the summer. **Nymphs** and adults dissuade **predators** with their **aposematic** color patterns. Adults also flash their stubby reddish wings, hiss from abdominal **spiracles,** kick with their spiny legs, and exude a malodorous and distasteful froth from the **thorax.** Some birds avoid exposure to this chemical defense by first removing the eastern lubber grasshopper's head and internal organs.

REPRODUCTION Females deposit 25–50 eggs in the soil, surrounding them in a protective **ootheca.** Nymphs hatch in about two weeks and, when abundant, damage flowers, vegetables, citrus, ornamental plants, and fruit trees. They sometimes climb into vegetation in groups, especially at night, presumably to avoid predation.

SIMILAR SPECIES Large, reddish brown and green, or mostly green and marked with brown, plains lubbers (*Brachystola magna*) are found in weedy fields and croplands in the Great Plains during summer and fall. Large and shiny black horse lubber grasshoppers (*Taeniopoda eques*) have yellow markings and live in desert shrub and oak habitats of southern Arizona and New Mexico.

Of Interest The Loggerhead Shrike, a North American songbird, impales these grasshoppers on plant thorns, apparently to let the chemicals dissipate, before eating them.

Nymph

- Size: L 52.0–65.0 mm
- Family: **Tettigoniidae**
- Life cycle: **One generation produced annually**
- Range: **Across southern and most of eastern United States**
- Food: **Adults eat leaves of broadleaf trees and shrubs**

Greater Angle-Wing Katydid

Microcentrum rhombifolium

*These katydids get their name because the upper margins of their uniformly green leaflike **tegmina** are distinctly bent, or angled, over the back.*

IDENTIFICATION The adults are green with hind **femora** that do not reach the last quarter of the tegmina. The hind wings extend just beyond the tips of the **tegmina.** Mature females have a flat curved **ovipositor.**

HABITS Males and females locate one another by producing sounds. Calling males produce loud lisps that are repeated every few seconds to attract distant females and a series of ticks resembling a thumbnail run slowly over the teeth of a comb when approaching a female ticking in response nearby. Both sexes hear one another with the aid of clearly visible earlike organs that open on the bases of their front **tibiae.** Although their calls are not particularly intrusive, large greater angle-wing katydids rarely go unnoticed when they are attracted to lights on warm evenings. Adult activity peaks in summer and fall across most of their range, but they are found year-round in Florida.

REPRODUCTION Flat, seedlike eggs are attached in a single row along leaf margins.

SIMILAR SPECIES The greater angle-wing katydid is the largest and most widespread of the six species of *Microcentrum* that occur north of Mexico. Male lesser angle-wing katydids *(M. retinerve)*, a widespread southeastern species, have a brown spot at the base of the wings. The California angle-wing katydid, *M. californicum*, occurs throughout most of California and southern Arizona.

Of Interest As adults, these katydids have the ability to fly if the need arises, but more often they move around by walking slowly and blending in with the leaves.

Eggs on branch

- Size: **L 15.0—25.0 mm**
- Family: **Gryllidae**
- Life cycle: **One generation produced annually**
- Range: **Across southern Canada and most of United States**
- Food: **Adults eat plant and animal tissues**

Fall Field Cricket

Gryllus pennsylvanicus

The fall field cricket is the most widespread species of Gryllus *in North America.*

IDENTIFICATION These robust, mostly dark brown or black crickets sometimes have a reddish tint. Their big, round heads bear long, slender **antennae** that are longer than the body. In males, the **tegmina** nearly cover the **abdomen,** but in females they are shorter. Both sexes have a pair of long abdominal **cerci** on the tip, but only the female has a needlelike **ovipositor.**

HABITS Males call to females from the entrances of their burrows, cracks on the soil, and other protected places on warm evenings in late summer and fall. Their calls consist of a series of two to three chirps per second. Although soothing, the amorous songs of male fall field crickets are sometimes annoying when they emanate from inside

walls and under appliances. Females appear to prefer the calls of older rather than younger males. Adults can damage crops but also eat eggs and pupae of insect pests.

REPRODUCTION Females deposit up to 50 eggs at a time in the soil and lay up to 400 eggs in a lifetime. Overwintered eggs begin hatching in May, and the **nymphs** reach maturity in late July and August.

SIMILAR SPECIES *Gryllus* was once thought to include a single species. Careful study of their calls and life cycles revealed 10 species in eastern North America and at least twice as many in the West, most of which await formal scientific description by entomologists.

> *Of Interest* Fall field crickets have been known to eat the seeds of several common weeds: crabgrass, ragweed, lamb's-quarters, and English plantain.

Adult female

- Size: **L 13.0–18.0 mm**
- Family: **Gryllidae**
- Life cycle: **Several generations produced annually**
- Range: **Across southern United States**
- Food: **Adults eat plant and animal tissues**

Tropical House Cricket
Gryllodes sigillatus

Also known as Indian house or banded crickets, tropical house crickets are probably native to southwestern Asia but now widespread in warm, mostly tropical regions around the world.

IDENTIFICATION These somewhat flattened, light yellowish-brown crickets have dark bands between the eyes and the front and rear margins of the **pronotum.** Both females and **nymphs** also have a broad band across the base of the **abdomen.** The **tegmina** cover half of the male's abdomen but are greatly reduced in size in females. Only females have **ovipositors.**

HABITS Tropical house crickets are found year-round hiding in rock piles, storm sewers, and crevices in walls, sidewalks, and other paved areas. They emerge at night to search for food and mates. They generally cause no harm indoors, but the incessant calls of the males can be very annoying.

REPRODUCTION Males attract females with a series of brief, high-pitched chirps and produce **spermatophores,** which both nourish and fertilize females. Eggs are deposited in the soil and develop into adults in about two or three months.

SIMILAR SPECIES House crickets (*Acheta domesticus*) have two distinct bands across the head, symmetrical blotches on the pronotum, and fully developed wings in both sexes. Widely sold as fish bait and animal food, this Old World cricket is established throughout the eastern United States and coastal southern California.

Of Interest These crickets are raised commercially as food for birds, reptiles, amphibians, and predatory arthropods kept as pets.

Adult female

- Size: L 13.0–14.0 mm
- Family: **Gryllidae**
- Life cycle: **One or two generations produced annually**
- Range: **Across southern Canada and United States**
- Food: **Adults eat many kinds of herbaceous plants**

Four-Spotted Tree Cricket

Oecanthus quadripunctatus

Four-spotted tree crickets are abundant in weedy fields, where the males are heard producing their steady trill during the day and late afternoon and into the night.

IDENTIFICATION These delicate and uniformly pale green insects have mouthparts that are directed forward, slender hind **femora,** and wings that are folded flat over the body. The **antennae** are long and slender and usually have four marks on the underside of the first two **antennomeres.** The **tegmina** are transparent, with green or greenish-yellow veins.

HABITS The call of the male four-spotted tree cricket is composed of a continuous trill rather than a series of chirps.

REPRODUCTION The yellow eggs are laid in loose rows within the stems of

small, pithy weeds and other herbaceous plants, especially wild carrot, in the eastern United States.

SIMILAR SPECIES Most of the 14 species of *Oecanthus* in North America, all commonly known as green tree crickets, are identified by the markings on the undersides of the first two antennomeres. Another widespread species, the snowy tree cricket (*O. fultoni*) is famous because it can be used to determine the temperature. In the East, count the number of chirps in 13 seconds and add 41 to come up with the temperature in degrees Fahrenheit. In the West, these crickets chirp slightly faster, and the formula varies throughout the region. For example, in Oregon, count the number of chirps in 12 seconds and add 38.

Of Interest During and after mating, female tree crickets sip a nutritious fluid from a gland located beneath the male's wing bases.

Nymph

- Size: **L 14.0–16.00 mm**
- Family: **Rhaphidiophoridae**
- Life cycle: **One or more generations produced annually**
- Range: **Eastern North America**
- Food: **Adults eat plant materials and small insects**

Greenhouse Stone Cricket
Diastrammena asynamora

*Greenhouse stone crickets, commonly found in basements and crawl spaces, are also known as spider crickets because their hairlike **antennae** and six gangly legs suggest the eight legs of an arachnid.*

IDENTIFICATION These brownish-yellow wingless crickets are mottled with brown. They have powerful jumping legs with smooth, not spiny, **tibiae** that can propel them up to four feet into the air.

HABITS Steady supplies of food, water, and shelter provided by human habitations guarantee the ubiquitous greenhouse stone crickets with everything they need. Native to Asia, greenhouse stone crickets first became established in greenhouses in North America and Europe, and they are now nearly cosmopolitan.

Outdoors, they hide under stones, logs, or among piles of firewood. Ivy and other low-growing ground cover also provide excellent hiding places. During inclement weather, these **nocturnal** insects seek shelter in garages, sheds, and basements by the dozens or even sometimes by the hundreds. Once indoors, they prefer resting on the walls of dark, humid spaces, such as bathrooms and laundry rooms. Clothing and linens may be damaged if the crickets cannot find suitable plant foods nearby.

REPRODUCTION Adult females have a conspicuous swordlike **ovipositor** on the tip of the **abdomen** through which they lay up to several hundred eggs in the soil.

SIMILAR SPECIES The decidedly domesticated greenhouse stone crickets resemble native species of camel crickets in the genus *Ceuthophilus*. Camel crickets are more robust and have distinctly spiny hind tibiae. They prefer living in tree hollows, cave entrances, cliff faces, and occasionally homes across the continent.

Adult female

- Size: **L 21.0–70.0 mm**
- Family: **Stenopelmatidae**
- Life cycle: **One generation produced every two to five years**
- Range: **Western North America, especially California**
- Food: **Adults eat both plant materials and small invertebrates**

Jerusalem Crickets
Stenopelmatus species

Jerusalem crickets look like a cross between Jiminy Cricket and the old Cootie toy, and they needlessly inspire fear.

IDENTIFICATION Also known as the potato bug or *niña de la tierra* (child of the earth) because of its head—outsize, smooth, shiny, and babylike—this large, brownish, and wingless cricket has a short **thorax** and a large, soft, and banded **abdomen.** The stout, spiny legs are adapted for burrowing through the soil.

HABITS Jerusalem crickets are usually found under objects on the ground or in burrows in gardens and parks in coastal plain, valley, foothill, mountain, and desert habitats. Although not aggressive, the adults are quite capable of defending themselves with their spiny legs and powerful **mandibles,** but they are not venomous. Adults are active on the surface during the cooler and wetter months in fall and winter and communicate with one another by drumming their abdomens on the soil. Dead adults in or near pools and streams are likely parasitized by long, slender horsehair worms.

REPRODUCTION Males and females grapple like wrestlers before mating, after which the female may kill and eat the male. The complete life history is unknown. Eggs are probably laid in small clutches in the soil. They hatch in spring and reach maturity in two to five years.

SIMILAR SPECIES The genus *Stenopelmatus* has about 60–80 species in North America, most of which have yet to be formally described by entomologists. Coastal and desert dune-dwelling species of *Ammopelmatus* are also called Jerusalem crickets.

Of Interest The common name of these distinctive insects probably originated when those startled by their unusual appearance declared, "Jerusalem!"

Close-up of head showing eyes and mandibles (mouthparts)

SINGING INSECTS

Two North American orders include singing insects: Orthoptera (crickets, katydids, grasshoppers, and their kin) and Hemiptera (cicadas). Orthopterans typically produce their calls by rubbing one body part with a sharp-edged "scraper" against another body part possessing a bumpy ridge or "file," a process known as **stridulation.** Males stridulate to attract mates. In some species, females respond with their own call.

CRICKET & KATYDID SONG

Male crickets and katydids use stridulatory organs located at the base of their **tegmina** and produce calls ranging from faint tinkling and pleasant chirping to loud, sustained buzzing or high-pitched rasping sounds. Amorous field crickets raise their tegmina slightly as they stridulate, while tree crickets raise theirs high overhead to help amplify their calls. Katydids barely move their wings at all as they make sound.

The cicada's song, often heard on warm summer days, comes from sound-producing organs called tymbals—the two small projections shown at the bottom of this photo.

Broad-winged katydids sing a five-syllabled lispy song during the day and a longer, more buzzy song at night.

GRASSHOPPER SONG

Many grasshopper species **crepitate,** or produce a clacking or rasping sound by rapidly rubbing filelike teeth on the inside of their hind **femora** against a ridge along the edge of the **tegmen.** Some species crepitate as they walk on the ground, while others do it in flight, sometimes flashing brightly colored hind wings.

CICADA SONG

Unlike singing orthopterans, male cicadas do not stridulate. Instead they rely on a pair of ribbed sound-producing organs under the base of the **abdomen** called **tymbals.** The convex and flexible tymbals are alternately buckled by high-speed muscle contractions to produce a rapid-fire series of clicks. The sound is directed through expanded **trachea** that function as resonating chambers inside the male's abdomen, which is largely devoid of other internal organs. Unlike those of orthopterans, the loud calls of cicadas can be heard at considerable distances.

A close-up of the stridulatory pegs on the inner surface of the grasshopper hind leg. These filelike teeth are scraped across a hard ridge along the edge of the tegmen.

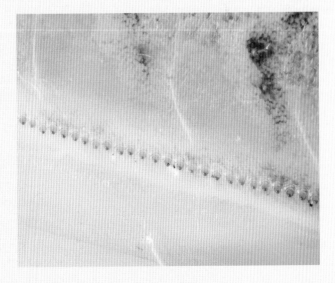

Stick Insects

Order Phasmida

Stick insects, or walking sticks, have legs that are all similar in form. They have chewing mouthparts and are herbivorous. All but one North American species are wingless. Development is by **hemimetaboly.**

- Size: L 55.0–101.0 mm
- Family: Diapheromeridae
- Life cycle: One generation produced annually
- Range: Eastern North America
- Food: Adults and nymphs eat leaves of deciduous trees and shrubs

Northern Stick Insect

Diapheromera femorata

*Northern stick insects are seldom seen in large numbers because of their **cryptic** colors and habits.*

IDENTIFICATION The uniformly green or brownish adults and **nymphs** often remain motionless with their front legs and **antennae** outstretched along a twig. They are easier to spot at night with a flashlight. The head is slightly longer than wide, with long antennae. The legs of the wingless adults are mostly long and slender, but the banded middle **femora** are somewhat swollen and each hind femur has a spine underneath.

HABITS These stick insects move slowly among trees and shrubs as they feed, consuming all the tissue of the leaf except for the thick veins at the base. Only rarely do they harm fruit and other broadleaf trees.

REPRODUCTION Males grasp females with abdominal claspers during copulation. Females haphazardly scatter up to 100 shiny black-and-gray-striped seedlike eggs onto the leaf litter. Nymphs hatch in spring and begin feeding on low-growing plants such as roses, strawberry, and blueberry. Larger nymphs move up into trees and appear to prefer oaks, basswood, and cherry. Five **molts** are required to reach maturity.

SIMILAR SPECIES With 10 species distributed east of the Rockies and in the Southwest, *Diapheromera* is the largest genus of stick insects north of Mexico.

Mating pair

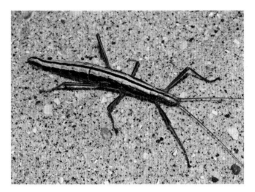

■ Size: L 40.0–80.0 mm
■ Family: **Pseudophasmatidae**
■ Life cycle: **One generation produced annually**
■ Range: **South Carolina to Florida and Texas**
■ Food: **Adults and nymphs eat leaves of deciduous trees and shrubs**

Two-Striped Stick Insect

Anisomorpha buprestoides

Large numbers of two-striped stick insects sometimes gather under loose bark or in other sheltered places.

IDENTIFICATION Brown and thick-bodied, the two-striped stick insects have two distinct pale stripes down their backs.

HABITS These stick insects eat the leaves of trees and shrubs, including oak, rose, rosemary, lyonia, and crepe myrtle, but they are not plant pests. Smaller male two-striped stick insects are often found riding on top of much larger females, especially in fall. Captive individuals are successfully maintained on privet. Threatened adults and **nymphs** squirt a strong-smelling milky spray from two **thoracic** glands located just behind the head that contains a single active component, anisomorphal—a chemical compound that is also produced by a mint plant and is closely related to the cat-stimulating ingredient found in catnip. These stick insects can accurately aim this defensive spray of anisomorphal up to 30 cm away to thwart attacks by birds, ants, beetles, and mice.

REPRODUCTION After mating, females dig small pits in sandy soil and deposit up to 10 eggs before covering them with sand. Females may also deposit their eggs at night in the crevices of pine bark.

SIMILAR SPECIES The smaller and paler *Anisomorpha ferruginea* lacks distinctive stripes and occurs throughout most of the southeastern United States, except Florida.

Of Interest Incidents have been reported of stick insects spraying humans and dogs in the eyes, resulting in severe pain and temporary blindness. Even inhalation of the spray can cause considerable discomfort.

Mating pair

- Size: L 65.0–90.0 mm
- Family: Heteronemiidae
- Life cycle: One generation produced annually
- Range: Oregon and California to Utah and Texas
- Food: Adults eat various scrubby plants and grasses

Western Short-Horned Stick Insect
Parabacillus hesperus

Western short-horned stick insects are the most commonly encountered and widely distributed native stick insect in the region, yet little is known about their natural history.

IDENTIFICATION *Parabacillus* species are easily distinguished from other stick insects in western North America by their short **antennae,** no longer than half the length of the front **femora.** They are very long, slender, wingless, and usually straw colored, but other color forms may vary from light to dark brown, sometimes with a pinkish hue. The last abdominal segment of the smaller and more slender male is about as wide as it is long.

HABITS Adult western short-horned stick insects are sometimes encountered resting on walls and window screens with their middle and hind legs outstretched and front legs extending forward. Adults appear to be most common in late summer and early fall in mountain and foothill areas, especially in chaparral habitats. Their food plants include rabbitbrush, burroweed, globe mallow, mountain mahogany, and buckwheat.

REPRODUCTION Females deposit dark, slender eggs resembling rye seeds at the bases of grasses. Some populations may be **parthenogenetic,** or capable of reproducing without mating. The population of individuals of Santa Cruz Island off the southern California coast is distinctly smaller than mainland populations.

SIMILAR SPECIES The Colorado short-horned stick insect (*P. coloradus*) occurs from Wyoming and Arizona to South Dakota and Texas. It also has short antennae, sometimes with a weakly striped head. The heavy-bodied female looks like *P. hesperus,* but the slender male's last abdominal segment is about twice as long as wide.

Seedlike eggs

True Bugs & Allies

Order Hemiptera

Hemipterans have piercing-sucking mouthparts. Most feed on plant sap, while others prey on insects or suck the blood of vertebrates. Adults usually have two pairs of wings. Development is by **hemimetaboly.**

- Size: **L 13.0–16.0 mm**
- Family: **Notonectidae**
- Life cycle: **One generation produced annually**
- Range: **Western North America to Dakotas**
- Food: **Adults eat small aquatic insects and crustaceans**

Kirby's Backswimmer
Notonecta kirbyi

Found year-round, this species is the largest and most widespread backswimmer found summer to fall in the West, most abundant from midsummer through fall in ponds, streams, and the occasional swimming pool.

IDENTIFICATION The greenish-yellow head of this deep-bodied insect is two-thirds the width of the **pronotum,** with red eyes separated on top and a black **scutellum.** The variable **hemelytra** are tan and black or red and mottled with black, each tipped with a small light spot. The fourth abdominal segment has a keel, or ridge, underneath that is bare, although the sides are covered with short, stiff hairlike **setae.**

HABITS These backswimmers propel themselves through the water with long, flattened, oarlike legs fringed with setae. Jerky swimmers, they must work to maintain their position in the water column. At rest, they hang upside down from the surface film with hind legs outstretched, ready to dart forward to capture hapless insects trapped on the surface or swimming nearby. To replenish their air supply, backswimmers use the stiff setae on the tip of the **abdomen** to break the surface film and draw a silvery bubble down their back.

REPRODUCTION Eggs are typically laid on aquatic vegetation or on submerged rocks and debris.

SIMILAR SPECIES Kirby's backswimmer resembles the eastern *N. undulata,* which has black eyes and a line of setae on the keel of the fourth abdominal segment.

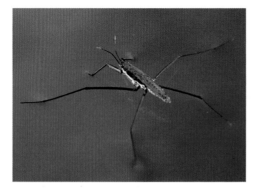

- Size: **L 9.0–17.0 mm**
- Family: **Gerridae**
- Life cycle: **One or two generations produced annually**
- Range: **North America**
- Food: **Adults prey on insects**

Water Strider
Aquarius remigis

Also known as pond skaters and Jesus bugs, the relatively large and robust Aquarius remigis *is one the best known and most widely distributed water striders in North America.*

IDENTIFICATION They are dark brown with a dull **pronotum** marked with a partial stripe down the midline. The first **antennomere** is slightly longer than the combined lengths of the second and third antennomeres. The **femora** of the front legs are uniformly dark, and the **tibia** of the hind leg is four times as long as the first **tarsomere.** The prominent spines flanking the tip of the **abdomen** curve slightly outward.

HABITS Water striders prefer slow-moving and permanent streams with coarse substrates, but they can also be found on streams, ponds, lakes, stock tanks, and sometimes even in swimming pools. They use their short forelegs to capture or scavenge insects trapped on the water's surface before piercing them with their sucking mouthparts. Individuals are occasionally seen resting on steeply sloping rocks and boards close to the water. In fall, water striders leave the water surface, and they over-winter under nearby rocks and leaf litter on land.

REPRODUCTION Females lay eggs just below the waterline. The eggs hatch in about two weeks, and then **nymphs** reach adulthood in less than two months. Adults and all five nymphal **instars** are found together on the water in summer. Adults within a population exhibit vary-ing degrees of wing development. Wing-less forms are typical, but local populations may have individuals with either reduced or fully developed wings.

SIMILAR SPECIES There are 46 species of genus *Aquarius* found in North America.

Wingless form

WALKING ON WATER

Water striders deftly glide over the surface of calm waters in search of both insect prey and potential mates—all without sinking. How do they do it? It's all about surface tension, long legs, and hairy feet.

The splayed-out legs of water striders evenly distribute their weight on the water. Each foot is kept dry by waterproof microsetae.

LONG LEGS & HAIRY FEET

The membranelike quality of the water's surface, or surface tension, is the result of water molecules at the air-water surface interlocking with one another. Water striders take advantage of surface tension. They splay out their six long legs, each foot resting inside a small surface dimple, to evenly distribute their weight and thus avoid sinking.

Each foot is densely coated underneath with tiny, waxy, needlelike microsetae. A **microseta** is less than two thousandths of an inch long, and a bundle of 25–30 microsetae approach the thickness of a human hair. The surface of each microseta is scored with tiny grooves that trap air to keep their feet dry.

These tiny water-repellant structures have inspired innovative new technologies for developing waterproof surfaces that reduce drag and increase buoyancy.

The swirls of dye trailing behind the water strider show the dual vortices, or water currents, under the middle and hind feet that propel the insect forward.

RIDING THE WAVE

Water striders are so buoyant they can support up to 15 times their own weight without sinking. They use their short front legs to locate and capture prey and to find mates. The longer middle and hind legs are used for propelling and steering, respectively. Water striders scull their middle legs like oars on a boat, simultaneously moving them backward and downward. This creates dual vortices, or water currents, beneath the dimples surrounding their feet that propel the insects forward.

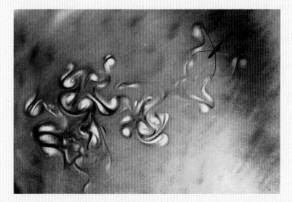

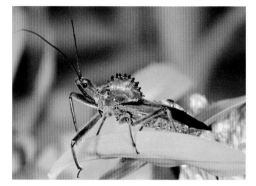

- Size: L 26.0–36.0 mm
- Family: Reduviidae
- Life cycle: One generation produced annually
- Range: Ontario to Florida, west to Iowa, Kansas, and New Mexico.
- Food: Adults prey on insects

Wheel Bug
Arilus cristatus

The wheel bug is one of the largest terrestrial bugs in North America, eye-catching for its shape as well as its size.

IDENTIFICATION This insect gets its name from the semicircular crest on the adult's **pronotum,** bearing 8–12 short, thick spines that resemble the gears of a wheel. The uniformly blackish-brown body is densely clothed with short, silky gray **setae,** while the **antennae, tibiae, tarsi,** and mouthparts are reddish brown. The enlarged front legs are used in conjunction with their thick piercing-sucking mouthparts for capturing insect prey.

HABITS Wheel bugs in all of their life stages can be commonly found on trees, shrubs, and other vegetation. There they hunt for **caterpillars,** beetles, and bees; adult wheel bugs will also attack brown marmorated stink bugs. Adults are typically encountered beginning in the late summer.

REPRODUCTION In fall, females glue somewhat hexagonal clutches of more than 100 dark brown to black, bottle-shaped eggs on twigs, tree trunks, and smaller branches. These eggs will hatch the following spring season, unless disturbed: At least two species of tiny wasps commonly parasitize wheel bug eggs. The antlike **nymphs** are black and run with their bright-red **abdomens** held high. Older nymphs are less ant-like and have gray and orange-red markings on their abdomens. The fifth and final nymphal stage resembles the adult but, like all nymphs, lacks fully developed wings.

SIMILAR SPECIES Of the five species placed in the mostly neotropical genus *Arilus,* only one occurs in North America. Wheel bugs and other assassin bugs are strictly insect **predators** and not **vectors** of disease.

Of Interest Both adults and nymphs of this species can deliver a painful bite if carelessly handled.

Nymph

- Size: **L 10.6–17.0 mm**
- Family: **Reduviidae**
- Life cycle: **One generation produced annually**
- Range: **North America**
- Food: **Adults prey on insects**

Assassin Bugs
Zelus species

Zelus assassin bugs are among the most commonly encountered of their kind in backyards and gardens.

IDENTIFICATION The long, slender adults are variably green, yellow, or brown with gray, red, and black markings, sometimes with a waxy coating. The elongate head bears a pair of long, slender **antennae** and long piercing-sucking mouthparts. The front and back corners of the **pronotum** are often armed with a conical spine. All of the legs are long and slender. The last abdominal segment of the male is cup shaped, while that of the female is more or less flattened.

HABITS The predatory adults and **nymphs** hunt for a wide variety of insects among grasses, shrubs, and low trees in gardens and parks. The forelegs are coated with a sticky glandular substance that helps both adults and nymphs to capture prey.

Nymph

REPRODUCTION Eggs are typically attached to the underside of leaves. The nymphs are usually bright green.

SIMILAR SPECIES Nine species of *Zelus* occur across North America, the most distinctive of which is the southeastern black and orange *Z. longipes*. The variable and widely distributed *Z. tetracanthus* is dusty brown, sometimes with black-and-white markings, and has four conical spines on the pronotum. The mostly western *Z. renardii* appears to thrive in disturbed habitats, including urban environments and agricultural systems. Now established in Chile and the Mediterranean via shipments of nursery stock, this species has become an important **predator** of sugarcane pests in Hawaii.

Of Interest Since they prey on other species that are agricultural pests, Zelus assassin bugs are considered beneficial insects when they inhabit fields of cotton, soybeans, alfalfa, and fruit tree orchards.

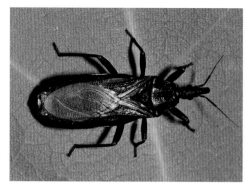

- Size: **L 13.0–23.0 mm**
- Family: **Reduviidae**
- Life cycle: **One generation produced annually**
- Range: **California to Colorado and Texas**
- Food: **Adults feed on the blood of wood rats**

Western Conenose
Triatoma protracta

Western conenoses belong to a group of insects known as kissing bugs, because they occasionally bite people on the face as they sleep, often on the eyelids or lips.

IDENTIFICATION Western conenoses are uniformly blackish brown to black with brownish to black **hemelytra** and light colored marks on the flared sides of the oval **abdomen.** A straight, three-segmented beak and **antennae** with seven **antennomeres** are borne on an elongate head.

HABITS Western conenoses are most active during late spring and summer and prefer living outdoors in rodent nests, where the **nocturnal** nymphs and adults prey mostly on wood rats. They occasionally suck blood from bedbugs and other blood-feeding insects, as well as other mammals. They seldom bite humans, probably only coming indoors when attracted to lights. A natural **vector** of the **parasite** that causes Chagas' disease, western conenoses don't transmit these disease-causing parasites to humans, although the saliva injected as they bite may cause anaphylactic shock in some sensitized people.

SIMILAR SPECIES It is unlikely that the 12 species of *Triatoma* in North America can transmit Chagas' disease to humans, because they are sylvatic and prefer to live and feed outdoors. Further, the disease-causing parasites are passed not through a bite but rather by the bug's **feces** when rubbed into the itching bite wound. Kissing bugs in the tropics defecate as they feed, but North American *Triatoma* defecate only after they have left the host. *Triatoma sanguisuga* is widespread in the East.

Nymph

Of Interest Western conenoses lay eggs in the summer and fall and overwinter as developing **nymphs,** emerging in adult form in the spring.

- Size: L 10.0–18.0 mm
- Family: Lygaeidae
- Life cycle: Up to three generations produced annually
- Range: Southeastern Canada and throughout United States
- Food: Adults feed on seeds and seedpods of milkweed

Large Milkweed Bug
Oncopeltus fasciatus

Large milkweed bugs are found in warmer months feeding and mating on the flowers and seedpods of milkweeds and their relatives growing in gardens, parks, old fields, agricultural edges, and woodland openings.

IDENTIFICATION The contrasting orange-red and black colors of feeding aggregations of large milkweed bugs and their **nymphs** serve as an **aposematic** billboard, warning potential vertebrate and invertebrate **predators** of their bitter flavor, acquired from the milkweed plant, their sole food. Mostly dark brown or black, these bugs have a Y-shaped mark on the head, margins of the **pronotum,** and two broad bands across the **hemelytra** that are all orange. Adult males have a black band across the underside of the fourth abdominal segment, while females have two black spots.

HABITS Adults tend to develop flight muscles in response to lack of food, but, when food is abundant, females lose flight muscle mass, rendering them flightless.

REPRODUCTION They have evolved adaptations that enhance their dispersal or increase their reproductive rates in response to season and available food. Up to 50 orange to orange-red eggs are laid in batches on leaves and stem. The nymphs pass through five **instars** in about a month before reaching adulthood.

SIMILAR SPECIES There are seven, possibly eight species of *Oncopeltus* found in North America, four or five of which occur in Florida. Other similar species include the smaller and more slender milkweed assassin bug *(Zelus longipes)* and the small milkweed bug *(Lygaeus kalmii).*

Nymph

Of Interest Hardy and easy to maintain in captivity, these bugs are commonly used as research animals in physiology experiments.

- Size: **L 10.0–12.0 mm**
- Family: **Lygaeidae**
- Life cycle: **Two or more generations produced annually**
- Range: **North America**
- Food: **Adults suck sap from a wide variety of weedy plants**

Small Milkweed Bug

Lygaeus kalmii

The common name of this species is a misnomer, because they feed not only on milkweeds but also on other weedy plants, including yarrows, ragworts, and other composites.

IDENTIFICATION These distinctive red-and-black bugs have a red spot at the base of the head and a red "X" on the back—spots that almost meet in the middle, formed by markings on the thick bases of the **hemelytra.** The membranous wing tips are either black (in the eastern populations) or black with white borders and a pair of whitish or grayish spots (in the western populations).

HABITS Overwintering adults emerge in spring and are found singly or in mating pairs throughout the summer. Subsequent generations of **nymphs** and adults will feed on milkweed, if available. Bugs feeding on milkweed

sap are distasteful, while those feeding on other plants are not. They also suck fluids from various dead insects, including ants, honey bees, milkweed leaf beetles, and *L. kalmii* adults; and they will attack monarch **caterpillars,** monarch pupae, and the eggs of the swamp milkweed beetle.

REPRODUCTION The females of this species lay the first clutches of eggs in spring on the surfaces of dead leaves or inside the hollow stems of dead plants. Nymphs feed on insect carrion and on the seeds of shining pepperweed and other **forbs,** or herbaceous weeds.

SIMILAR SPECIES Small milkweed bugs are sometimes mistaken for large milkweed bugs and boxelder bugs, but neither of those species has the telltale red "X" on their backs. The false milkweed bug, *L. turcicus,* has a red "Y" on the head and virtually no white markings.

Nymph

Of Interest This species is an omnivore, known to feast on both plants and insects.

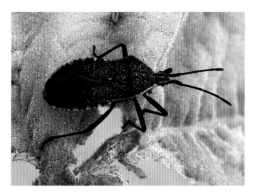

- Size: **L 13.0–18.0 mm**
- Family: **Coreidae**
- Life cycle: **Two or more generations produced annually**
- Range: **North America**
- Food: **Adults drink sap from squash and related plants**

Squash Bug
Anasa tristis

The bane of vegetable gardeners across North America, squash bugs attack various members of the squash family, including cucumbers and pumpkins.

IDENTIFICATION These "broad-shouldered" bugs are mostly dull grayish brown and covered with large, dark **punctures.** The dark brown head has a small **tubercle** behind the base of each **antenna** and three yellowish stripes. The **pronotum** and thickened, leathery bases of the **hemelytra** have brown speckles. The flared margins of the **abdomen** are alternately marked with somewhat rectangular brownish and yellowish patches.

HABITS Overwintering adults emerge in early summer. Both the adults and **nymphs** use their strawlike piercing-sucking mouthparts to draw sap from fruits, leaves, and stems. Initial feeding on these tissues produces small yellowish wounds, which are soon invaded by microorganisms that quickly decompose tissues, resulting in brown or wilted leaves and collapsed, rotting stems.

REPRODUCTION Females usually attach masses of coppery red eggs on the underside of leaves but will occasionally utilize stems. Newly hatched nymphs are pale green, eventually turning gray as they mature. Northern populations produce only one generation annually, but two or more may occur in the warmer South.

SIMILAR SPECIES There are six species of *Anasa* that occur north of Mexico. Of these, the horned squash bug, *A. armigera,* is sometimes a pest of squashes in the South.

Nymphs and eggs

Of Interest Companion planting may be able to keep squash bugs out of the garden. Flowers of nasturtium and tansy can work as natural repellents.

- Size: **L 16.0–20.0 mm**
- Family: **Coreidae**
- Life cycle: **One generation produced annually**
- Range: **Western North America and Atlantic states**
- Food: **Adults feed on conifer seeds**

Western Conifer Seed Bug
Leptoglossus occidentalis

Western conifer seed bugs belong to a family of insects commonly called leaf-footed bugs because of the flattened expansions on the hind **tibiae** *of many species.*

IDENTIFICATION They are dull brown with a faint zigzag band across the bases of the **hemelytra.** The flared abdominal margins have alternating light and dark rectangles.

HABITS Overwintering adults emerge in spring and seek coniferous trees on which to feed, especially various species of pines and Douglas fir. The adults feed on ripening conifer seeds and fly with a distinctive buzzing sound, revealing bright–yellowish orange areas on the **abdomen.** They sometimes become a nuisance in fall when they enter homes and out buildings through vents, torn screens, or gaps near windows, electrical conduits, and plumbing to avoid cold weather. Western conifer seed bugs do not bite, lay their eggs indoors, or spread disease—but they will release a disagreeable odor when disturbed.

REPRODUCTION Females lay rows of eggs on pine needles; they hatch within two weeks. Developing orange and brown to reddish-brown **nymphs** feed on soft cone scales and needles, reaching maturity in late summer.

SIMILAR SPECIES The genus *Leptoglossus* includes 11 species of leaf-footed bugs north of Mexico, 3 of which have faint or bold zigzag bands across the hemelytra: *L. brevirostris, L. clypealis,* and *L. zonatus.* Western conifer seed bugs are distinguished by pointed, not spinelike, heads with long mouthparts reaching the bases of their middle legs and rounded rather than scalloped hind tibial expansions.

Of Interest Since these bugs enter homes for the winter, the best control measures include sealing off the access points in late summer, before they try to get in.

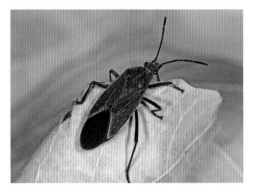

- Size: L 11.0–14.0 mm
- Family: Rhopalidae
- Life cycle: Two generations produced annually
- Range: Eastern North America
- Food: Adults suck sap mostly from boxelders

Boxelder Bug
Boisea trivittata

Masses of red-and-black boxelder bugs on tree trunks, fences, and buildings are a familiar sight on warm, sunny days in spring and fall—seldom more than a nuisance, these aggregations never warrant control measures.

IDENTIFICATION They are black with conspicuous red eyes and three red stripes on the **pronotum,** with red margins on the thickened bases of the **hemelytra** and **abdomen.**

HABITS Overwintering adults emerge in spring, and females soon being laying eggs. Both **nymphs** and adults drink sap from leaves, flowers, and seeds, but they seldom damage trees. They will occasionally feed on developing fruits, causing puckered "cat facing" injuries. With the arrival of cooler fall temperatures, adults seek warmth by congregating on sunny walls and porches before invading homes through windows, ducts, and other gaps to escape frost.

REPRODUCTION Females lay batches of rusty brown eggs near fallen seeds of boxelder, maples, and occasionally ash. The young, mostly red nymphs feed on seeds and other low herbaceous growth; they will even scavenge dead insects and cannibalize other nymphs. Mature nymphs with black wing pads complete their development in early summer. Another generation is produced in late summer.

SIMILAR SPECIES The western boxelder bug, *B. rubrolineata*, occurs in the West and is similar in appearance. Another similar species, the red-shouldered or golden-tree bug *(Jadera haematoloma)*, occurs across the southern half of the United States. It is mostly grayish black with red eyes and red stripes on the sides of the pronotum.

Adults and nymphs

- Size: L 14.0–17.0 mm
- Family: Pentatomidae
- Life cycle: Two to five generations produced annually
- Range: North America
- Food: Adults suck juices from fruits

Brown Marmorated Stink Bug
Halyomorpha halys

Native to eastern Asia, the brown marmorated stink bug was first recorded in Pennsylvania in 1996 and is now widely established throughout the United States.

IDENTIFICATION This grayish-brown, shield-shaped stink bug is mottled with black and has a blunt head, alternating light and dark markings on the **antennae** and edges of the **abdomen,** and a distinct pale ring around each **tibia.**

HABITS Overwintering adults emerge in spring. Adults and **nymphs** attack a wide range of plants, including many ornamentals, vegetables, and fruit trees. Their feeding activities distort and scar fruits and vegetables and introduce microorganisms that cause decay. In fall, adults seek shelter inside homes and outbuildings, sometimes by the hundreds or thousands, via gaps in

vents, windows, and doors. Brown marmorated stink bugs do not bite people or pets, or lay their eggs indoors; and they are not **vectors** of disease, but they will release a disagreeable odor when crushed.

REPRODUCTION Females lay batches of 20–30 light green eggs on the undersides of leaves that hatch in about a month. The developing nymphs also have white bands on their antennae and legs. Up to five generations are produced annually in the South.

SIMILAR SPECIES Bark stink bugs in the genus *Brochymena* are similar in size and appearance, but the adults and nymphs lack the white bands on the antennae and legs.

Of Interest The indoor use of pesticides against brown marmorated stink bugs is discouraged because dead bugs in walls and other inaccessible areas lead to other insect pest infestations.

Nymph

- Size: L 13.0–19.0 mm
- Family: Pentatomidae
- Life cycle: One or two generations produced annually
- Range: Across southern Canada and United States
- Food: Adults feed on fruit trees, legumes, and other plants

Green Stink Bug

Chinavia hilare

Green stink bugs are among the most commonly encountered stink bugs in North America, especially in gardens, orchards, crop lands, and woodlands.

IDENTIFICATION The adult's bright-green body is narrowly bordered with yellow, orange, or red, while the last three **antennomeres** are black. Early **instar** nymphs are brightly colored, but they start turning green as they mature. Both adults and **nymphs** have large stink glands that open into a long channel on the underside of the **thorax** and through which they release large amounts of a foul-smelling liquid when disturbed.

HABITS Green stink bugs feed on more than 30 plant species, including apples, pears, green beans, soybeans, and other legumes. Adults prefer to feed on fruits and seeds in vegetable crops, but they will also attack the foliage and stems of fruit trees. Their feeding results in puckering at the bite wound called cat facing. In cotton fields, green stink bugs cause economically significant damage: They reduce harvestable cotton, reduce seed germination, and increase the volume of immature fibers.

REPRODUCTION Overwintered adults emerge in spring to feed and reproduce. One generation is produced in the North, with adults beginning to appear in mid-September and early October. Two generations are completed under favorable conditions in northern regions and in the southeastern and Gulf states.

SIMILAR SPECIES The green stink bug is similar to *Nezara viridula* but is distinguished by its black outermost antennomeres. *Acrosternum pennsylvanicum* is also similar, but the sides of its **pronotum,** just behind the head, are distinctly arched, while they are somewhat straight in green stink bugs. Several other genera of stink bugs include species that are green.

Nymph

- Size: L 8.0–12.0 mm
- Family: Pentatomidae
- Life cycle: One to four generations produced annually
- Range: Southern half of United States
- Food: Adults suck sap from mustards and related plants

Harlequin Bug
Murgantia histrionica

Harlequin bugs, also known as calico or fire bugs, are common sights in gardens, parks, vacant lots, and fields in late spring and summer.

IDENTIFICATION Marked in red, orange, yellow, white, and black, these variable stink bugs range from mostly black to mostly orange. The **nymphs** are similarly colored but more rounded in outline.

HABITS Overwintered adults emerge in spring and feed on wild mustard and other weedy members of the mustard family. By summer, they begin moving into gardens to feed on cabbage, radish, and mustard greens. All stages are major pests of cabbage, broccoli, radishes, and related crops. Using piercing-sucking mouthparts to suck sap, they cause plants to wilt, brown, and die. Adults

Eggs and nymph

migrate in spring and summer, sometimes reaching as far north as New England, but their northern limits vary annually depending on the severity of each winter. Adults overwinter in sheltered sites under plant debris.

REPRODUCTION They are often seen on food plants, copulating end-to-end. Females lay two rows of black-and-white, barrel-like eggs on their food plants that hatch in about a month. The nymphs take two or three months to mature, depending on temperature. Up to four generations are produced annually in the warmer South.

SIMILAR SPECIES The bagrada or painted bug, *Bagrada hilaris,* is an invasive species from Africa that has become established in southern California and southern Arizona. It has food plant preferences similar to that of harlequin bugs, but it is much smaller, and its **pronotum** and **scutellum** have only longitudinal markings.

- Size: L 27.0–30.0 mm
- Family: Cicadidae
- Life cycle: One generation produced every 17 years
- Range: Connecticut to Georgia, west to Wisconsin, Nebraska, and Kansas
- Food: Adults and nymphs suck tree sap

Seventeen-Year Cicada

Magicicada septendecim

The mass emergence of seventeen-year cicadas, heralded by an intense rising and falling chorus of thousands of males seeking a mate, is one of the great phenomena of the natural world.

IDENTIFICATION The black and red-eyed adults have orange wing veins and broad orange bands across the underside of each segment of the **abdomen.** The sides of the **thorax** are also marked with orange, just behind the eye and at the wing bases.

HABITS They spend most of their lives underground as **nymphs** sucking sap from roots but seldom harm healthy, mature trees. There are 12 different populations of seventeen-year cicadas, or broods, and each brood emerges in late spring every 17 years. A few individuals, or stragglers, emerge a few years earlier or later than expected.

REPRODUCTION After mating, females cut slits in living twigs with their reproductive organs, sometimes causing minor damage in tree nurseries. They lay an egg inside each slit. Hatching nymphs drop to the ground and burrow into the soil to begin feeding for the next 17 years.

SIMILAR SPECIES Two similar species of periodical cicadas emerge with *M. septendecim. Magicicada cassini* is smaller, lacks the orange spot behind the eye, and usually has a completely black abdomen underneath. The less common *Magicicada septendecula* is similar in size to *M. cassini* and lacks the spot behind the eye but has narrow bands across the underside of the abdomen. The male's call for each species is distinctive.

Of Interest Brood X is the largest brood of seventeen-year cicadas. Extending from Michigan and eastern Illinois to New Jersey and northern Georgia, this brood will reemerge in 2021.

Adult emerging from nymphal skeleton

- Size: L 34.0–46.0 mm
- Family: **Cicadidae**
- Life cycle: **One generation produced every 3–5 years**
- Range: **Extreme southern Canada and eastern United States**
- Food: **Adults and nymphs suck sap of oaks**

Northern Dusk-Singing Cicada
Neotibicen auletes

The largest annual cicada in eastern North America, male northern dusk-singing cicadas produce their loud, pulsating drones in early evening.

IDENTIFICATION Northern dusk-singing cicadas are black with dull green and brown markings, especially on the **thorax,** and a variable dusty coating of white wax. Their prominent eyes are tan or light green.

HABITS These cicadas favor suburbs, parks, oak woodlands, and deciduous forests in the piedmont and coastal plain. Although widespread east of the Mississippi River, this species is less common in southern Ontario and southern coastal habitats. Adults emerge in the summer, and males typically call from high up in the tree canopy. Both sexes are strong fliers, and they are occasionally attracted to light. When captured, these insects produce a loud alarm call to startle **predators.** Both sexes use their short, piercing-sucking mouthparts in order to draw fluids from the **xylem** of oak trees.

REPRODUCTION With their **ovipositors,** females cut slits into branches, into which they lay eggs that will hatch in about two weeks. The **nymphs** drop to the ground and burrow along the roots of oak trees to feed on sap. Mature nymphs emerge at night to crawl up tree trunks and other vertical surfaces to complete their transformation to adulthood. Although the nymphs take three to five years to develop, adults emerge annually because the generations overlap one another.

SIMILAR SPECIES Of the 27 species of *Neotibicen,* all but one occurs in central and eastern North America. The largest cicada in the West is *N. cultriformis.* With practice, most of these species can be distinguished by the male's call alone.

Adult emerging from nymphal skeleton

- Size: L 23.0–25.0 mm
- Family: Cicadidae
- Life cycle: One generation produced every 3–5 years
- Range: Southeastern California and southern Nevada to Colorado and Arizona
- Food: Adults and nymphs suck tree sap

Apache Cicada
Diceroprocta apache

The loud steady buzz of male apache cicadas fills the air both day and night at the height of summer in riparian and irrigated habitats across the lower Colorado River Basin.

IDENTIFICATION This dark brown or reddish-brown cicada has paler brownish markings on its head and **thorax,** and reddish eyes. The **pronotal** "collar" just in front of the wing bases is uniformly pale yellow, straw colored, or tan.

HABITS Adults begin emerging in June and reach their peak of activity in July. Like other insects, males typically emerge before the females. They feed on the **xylem** sap of a wide variety of native tree species, including willow, cottonwood, seepwillow, paloverde, and mesquite. They also utilize non-native species, such as salt cedar, or species found growing in urban and agricultural settings, such

as date palms and fruit trees. The xylem feeding activities of the **nymphs** may significantly increase soil moisture levels at the surface as a result of their urine output.

REPRODUCTION After mating, females deposit their eggs in small twigs of suitable trees. Although the nymphs take at least three years to develop, the adults emerge annually because the generations overlap one another.

SIMILAR SPECIES Most of the 21 species of *Diceroprocta* occur in western North America. *Diceroprocta vitripennis* is widespread in the central United States, however, while coastal *D. viridifascia* occurs in the southeastern and Gulf Coast states.

Of Interest Large summertime emergences of many species of cicadas provide **predators** with an important and abundant source of food. Mammals, birds, reptiles, and other arthropods all feed on cicadas.

Adult with wings spread

PERIODICAL & ANNUAL CICADAS

For many, calling cicadas provide the perfect soundtrack for summer. The amorous males produce their distinctive soft clicks, loud buzzes, shrill trills, and pulsating whines with a pair of **tymbals** located on the first abdominal segment (see "Singing Insects," pp. 78–79). Their songs serve primarily to attract the attention of females, and most species are easily identified by their call alone.

Periodical cicadas in the genus *Magicicada* emerge periodically every 13 or 17 years in late spring. They have red compound eyes and black bodies variously marked with orange.

Adult cicadas live only for a few weeks. Females deposit their eggs, one at a time, in twigs and grass stems. Depending on the species, the subterranean **nymphs** may spend up to 17 years underground, sucking plant sap from tree roots. Mature nymphs typically emerge at night and crawl up tree trunks and fences to complete their development, leaving behind their cast **exoskeletons,** or **exuviae,** commonly known as "shells."

Most adult cicadas are strong fliers and spend their short lives high in the trees, feeding on plant juices with their

Annual cicadas, such as this *Neotibicen*, do not have red compound eyes. Note the three simple eyes, or ocelli, on top of the head. The bodies of annual cicadas are variously black, green, or brown, sometimes with a pale waxy coating. They emerge annually in summer.

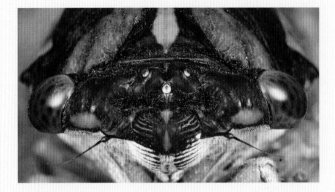

piercing-sucking mouthparts. A few species are regularly attracted to bright lights at night. North American cicadas fall into two broad categories depending on the timing of adult emergence: periodical and annual.

PERIODICAL CICADAS

Periodical cicadas, all in the genus *Magicicada*, have distinctly red eyes. Their bodies are mostly black with orange markings, legs, and wing veins. All seven species are restricted to the eastern United States, where they are distributed in 15 discrete populations or broods. Twelve of those broods (representing three species) emerge in late spring every 17 years, and three emerge every 13 years (four species). Stragglers may emerge a year before or after expected emergences. These massive emergences of periodical cicadas have led many people to refer to them mistakenly as "locusts."

ANNUAL CICADAS

The remaining 163 species of cicadas in North America, most of which are in the genera *Cacama, Diceroprocta, Neotibicen, Okanagana*, and *Platypedia*, emerge every year and are collectively referred to as annual cicadas. The eyes of annual cicadas are not red, and their bodies are variously black, green, and brown, sometimes with a pale waxy coating. Depending on the species, the nymphs may take as long as five years to develop. Unlike the synchronous regional mass emergences of periodical cicadas, the life cycles of individuals within a population of annual cicadas overlap one another so that adults emerge in midsummer every year.

- Size: L 9.0–13.0 mm
- Family: Membracidae
- Life cycle: Two generations produced annually
- Range: British Columbia, Pacific, southern, and Atlantic states
- Food: Adults and nymphs suck oak sap

Oak Treehopper
Platycotis vittata

A combination of age, sex, season, and geography determines the presence of horns and the color pattern of oak treehoppers.

IDENTIFICATION Robust and triangular oak treehopper adults may or may not have a flattened, forward projecting **pronotal** horn of variable length and short hind **tarsi.** **Teneral** adults are turquoise with red eyes and stripes, while mature individuals are grayish or brownish green overall with small orange or yellow spots. The red-eyed and horned or hornless **nymphs** are strikingly marked, with black-and-white stripes interspersed with red abdominal patches.

HABITS The bright **aposematic** colors of the nymphs and young adults warn **predators** of their distastefulness, while the mature, more palatable adults are **cryptic** to avoid the attention of predatory birds and lizards as they disperse to search for mates.

REPRODUCTION Overwintered females deposit their eggs in slits harmlessly cut in live twigs of deciduous and evergreen oaks. The nymphs hatch just about the time of bud break and, with their mothers, form **subsocial** aggregations with up to 100 individuals. Mothers stand guard over the nymphs, cutting slits in the bark to facilitate their feeding on **phloem** sap. Newly emerged adult females mate but don't begin laying eggs until the fall.

SIMILAR SPECIES In Florida, the oak treehopper is sometimes confused with another horned treehopper known as the thorn bug, *Umbonia crassicornis,* but the thorn bug's horn is decidedly more upright and its hind tarsi distinctly longer.

Of Interest The U-shaped map of the oak treehopper's distribution pattern embraces the mild winters along the Atlantic and Pacific coasts and across the warmer southern states.

Nymph

- Size: L 8.0–10.0 mm
- Family: **Aphrophoridae**
- Life cycle: **One generation produced annually**
- Range: **Widespread in North America, except in Southwest**
- Food: **Adult and nymphs suck sap from many plants**

Meadow Spittlebug
Philaenus spumarius

Native to Eurasia, meadow spittlebugs—also known as froghoppers—are widely established across most of North America.

IDENTIFICATION Variably marked adults are grayish green, yellowish, brown with pale or dark spots, or black with pale margins. In pale forms, the head has two black spots. In all color forms, the wings have distinctly raised veins. Females are slightly larger than males.

HABITS Adults appear in late spring and feed on a variety of plants throughout the summer, including alfalfa, red clover, wheat, oats, corn, and strawberries, although they are rarely pests. Meadow spittlebugs jump with the aid of elastic pads of a protein, called resilin, that stores muscle energy at the bases of their hind legs. Energy suddenly released to their feet propels them more than 100 body-lengths in a single bound.

REPRODUCTION Females lay masses of 50 or more eggs in late summer and fall, depositing them on plant stems near the ground. Overwintered eggs hatch in spring. The **nymph,** or spittlebug, is bright green with black **antennae** and surrounds itself in a white, frothy mass of bubbles extruded from the anus. This distinctive spittle is actually a mixture of waste and glandular secretions and protects them from **predators,** temperature fluctuations, and dehydration.

SIMILAR SPECIES The two-lined spittlebug, *Prosapia bicincta* (Cercopidae), is a conspicuous species in the East. Both nymphs and adults prefer feeding on warm season grasses. The distinct adults are dark brown with two narrow orange bands across the forewings.

Of Interest Scientists are studying the spittle production of these bugs to develop stronger and more efficient foam insulation.

Nymph exposed in spittle

- Size: L 11.0–13.0 mm
- Family: **Cicadellidae**
- Life cycle: **Two generations produced annually**
- Range: **Eastern United States to North Dakota and Texas**
- Food: **Adults and nymphs suck plant sap**

Broad-Headed Sharpshooter

Oncometopia orbona

"Sharpshooter" is the common name for some leafhoppers that filter nutrient-laden fluid and forcibly excrete it from their anus as a fine stream of droplets as they feed.

IDENTIFICATION Also known as blue sharpshooters, these insects have a short, convex head, round in front. The body and forewings are covered with scattered black markings. The head, front of the **pronotum, scutellum,** and legs are orange yellow, while the rest of the pronotum and forewings are blue. In the South, this species overwinters as an adult.

HABITS Adults feed singly or in groups on **phloem** sap of herbaceous plants growing in open fields all summer and then trees and shrubs in the fall. They are **vectors** of bacteria

that cause serious damage to fruit crops: phony peach in peaches, Pierce's disease in grapes, and oleander leaf scorch.

REPRODUCTION Female broad-headed sharpshooters have **brochosomes** on their forewings: These are small chalky lumps of intricately sculpted microscopic granules of protein that they apply to their eggs to protect them from drying out before hatching. Females lay eggs on herbaceous vegetation. **Nymphs** reach maturity in about two months.

SIMILAR SPECIES The less common *O. nigricans* is similar in size and color but with forewings that are lined rather than spotted. The southeastern glassy-winged sharpshooter *(Homalodisca vitripennis)*, accidentally introduced into California on shipments of ornamental plants, transmits several bacterial pathogens that infect citrus and many other plants. It is duller than either *Oncometopia* and has a longer, flatter, and more pointed head.

Brochosomes on wing

- Size: **L 1.0–2.0 mm**
- Family: **Aleyrodidae**
- Life cycle: **Several generations produced annually**
- Range: **North America**
- Food: **Adults and nymphs suck plant sap**

Greenhouse Whitefly

Trialeurodes vaporariorum

Greenhouse whiteflies are important pests of fruits, vegetables, and ornamental plants, especially in greenhouses.

IDENTIFICATION The pale-yellowish adults hold their four white and wax-coated wings almost parallel to the leaf surface.

HABITS Adults and **nymphs** use their short piercing-sucking mouthparts to extract **phloem** sap from leaf veins, often transmitting viral plant pathogens as they feed. They are particularly fond of vegetable crops related to cucumbers and tomatoes but also utilize woody ornamental shrubs and trees such as rose, ash, dogwood, sycamore, sweet gum, and honey locust.

REPRODUCTION The tiny, newly hatched leggy nymphs, or crawlers, are barely noticeable.

They soon transform into flat, pale, translucent, and legless insects fringed with short, fine wax filaments. Before reaching adulthood, the late-stage nymphs enter an immobile and non-feeding pupalike stage with distinctively fine wax filaments that radiate outward, giving them an almost spiny appearance. The number of wax filaments in those stages is determined by their population density and the "hairiness" of the host plant's leaf surface.

Of Interest Whiteflies are important pests of citrus and greenhouse plants. Heavy whitefly feeding results in disease transmission and also deprives plants of nutrients, discolors and dries out leaves, and covers leaf surfaces with a sticky **honeydew** that encourages the growth of black sooty mold and thereby reduces photosynthesis.

Adults with nymphs

- Size: L 2.0–3.0 mm
- Family: Aphididae
- Life cycle: Multiple generations produced annually
- Range: North America
- Food: Adults and nymphs suck on sap of oleander and milkweeds

Oleander Aphid
Aphis nerii

*The **adventive** oleander aphid, or milkweed aphid, originally from eastern Asia, is now a nearly cosmopolitan urban pest.*

IDENTIFICATION The small, soft-bodied, bright-yellow adult and **nymph** forms of the oleander aphid have black **antennae**, legs, **cornicles,** and tip of the **abdomen.** When a plant that they are feeding on becomes crowded or begins to die, forms with dark-veined wings appear.

HABITS These aphids infest oleander, milkweeds, and related plants. Among their aggregations, it is not unusual to find light brown and hollowed-out aphids called "mummies": dead aphids that have been parasitized by the larvae of **parasitoid** wasps. Although implicated in transmitting some plant pathogens, oleander aphids are considered pests primarily because of their unsightly colonies on infested plants, including the **honeydew** residue they produce, which encourages black sooty mold and blocks photosynthesis. Larvae of flies, lacewings, and lady beetles prey on aphids and often occur in or near the aphids' colonies. When attacked, these aphids collectively twitch their bodies, kick their legs, and secrete droplets of yellowish sticky wax from the tips of their cornicles. Parasitoid wasps are sometimes partially or completely immobilized by these droplets.

REPRODUCTION Males are unknown. Rather than laying eggs, the **partheno-genetic** females give live birth to genetically identical nymphs. Under ideal conditions, nymphs reach maturity in just a few weeks.

SIMILAR SPECIES Several Old World *Aphis* species have become pests in North America, including the soybean aphid (*A. glycine*), cotton aphid (*A. gossypii*), corn root aphid (*A. maidi-radicis*), apple aphid (*A. pomi*), and others.

Nymph

- Size: L 2.0–3.0 mm
- Family: Pseudococcidae
- Life cycle: Two or three generations produced annually outdoors
- Range: Across southern United States
- Food: Adult females and nymphs suck sap from many plants

Long-Tailed Mealybug
Pseudococcus longispinus

Long-tailed mealybugs are **adventive** *insects found on greenhouse and indoor plants year-round, but they may also occur outdoors, especially in southern states.*

IDENTIFICATION Adult females are soft, uniformly white, flattened, and oval with long waxy filaments protruding from the tips of their **abdomens,** a pair of which may exceed the length of the body. They are coated in waxy dust and ringed with shorter filaments. The males have two wings and a pair of long abdominal filaments. Rarely seen, they live only long enough to mate. Long-tailed mealybugs feed on many different kinds of flowers and vegetables, ground covers, ferns, palms, and shrubs. They are common on avocado trees in California. Infestations cause early leaf drop and coat leaf surfaces with sticky **honeydew** that fouls foliage and fruit and hinders photosynthesis.

Adult male

REPRODUCTION Females secrete themselves in stem crotches and lay eggs that hatch almost immediately. Young **nymphs** remain under their mother briefly before venturing away to suck **phloem** sap from leaves and small branches. They reach adulthood in about 45 days.

SIMILAR SPECIES The citrus mealybug *(P. citri)*, also an important greenhouse and indoor plant pest, is distinctly segmented and lacks the long abdominal filaments. A pest of native trees across southern Canada and northern United States, Comstock mealybugs *(P. comstocki)* attack fruit trees, including pear, apple, and peach.

Of Interest Various insects are important in controlling mealybug outbreaks, including green lacewings, lady beetles, and predatory bugs and fly larvae, as well as select **parasitoid** wasps that specifically attack mealybugs.

- Size: L 9.0–13.0 mm, with egg sac
- Family: Margarodidae
- Life cycle: Two to three generations produced annually
- Range: Southern California, Arizona, Gulf States north to Virginia
- Food: Adult females and nymphs suck plant sap

Cottony Cushion Scale
Icerya purchasii

First noticed in California in 1868 attacking acacias imported from Australia, cottony cushion scales nearly destroyed the state's fledgling citrus industry by 1885.

IDENTIFICATION The wingless yellowish, orange-red, or brown female is partially or completely covered with whitish or yellowish wax. Tiny reddish **haploid** males have two dark wings and are rare.

HABITS Females atop their white, waxy, and fluted egg sacs are conspicuous on a wide range of trees and shrubs. Infestations result in yellowing and premature leaf and fruit drop. The production of **honeydew** attracts ants and leads to the development of black sooty mold. Greenhouse infestations may occur throughout North America.

REPRODUCTION Some **diploid** females are capable of self-reproducing using sperm produced from tissues derived from their fathers. The orange, black-legged **nymphs,** sometimes coated with wax, are often seen crawling over adult females. In cooler weather, the entire life cycle takes about two or three months to complete, producing as many as three generations annually, but many more are possible in greenhouses and warmer areas.

SIMILAR SPECIES The cottony maple scale, *Pulvinaria innumerabilis* (Coccidae) produces soft, cottony egg sacs that resemble small marshmallows. Found throughout southern Canada and the United States, they attack many species of hardwoods.

Of Interest When *I. purchasii* infested California, entomologists went to Australia to bring back natural enemies, including the vedalia lady beetle *(Rodalia cardinalis).* Within a year, the beetle proved an effective **predator**—a success that marked the beginning of modern **biological control.**

Adult male

Lacewings, Antlions & Allies

Order Neuroptera

Neuropterans are soft-bodied insects with four finely veined membranous wings that are held rooflike at rest. Adult and larvae have chewing mouthparts and prey mostly on insects. Development is by **holometaboly.**

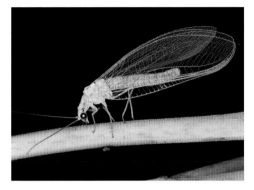

- Size: L 12.0–14.0 mm
- Family: Chrysopidae
- Life cycle: One or more generations produced annually
- Range: North America
- Food: Adults eat aphids, honeydew, and pollen; larvae prey on small insects

Green Lacewings

Chrysopa and Chrysoperla species

Active at dusk on warm evenings in spring and summer, green lacewings are a common sight at porch lights in cities and suburbs.

IDENTIFICATION The soft-bodied adults are bright green with long, thread-like **antennae** and cylindrical **abdomens.** They fold their clear, green-veined wings rooflike over the body when at rest.

HABITS These beneficial insects are commonly found in gardens, parks, and woodlands. Their eggs are sold as **biological controls** of plant pests because the larvae, dubbed aphid lions, have voracious appetites for aphids and other small insects. The slow-flying adults can hear ultrasonic signals of bats through earlike organs under their wing bases and release a foul-smelling fluid when attacked.

REPRODUCTION Females lay eggs singly or in batches on leaf surfaces, each affixed to a slender silken stalk, presumably to keep them out of reach of their canni-balistic siblings. The larvae emerge in about a week and immediately begin hunting for a wide range of small insect prey (aphids, **caterpillars,** beetles). The mostly light brown larvae superficially resemble lady beetle larvae, but the long, sicklelike **mandibles** easily distinguish them.

SIMILAR SPECIES There are nine species of *Chrysopa* and seven species of *Chrysoperla* in Canada and the United States. Some species are typically light brown.

Larva

- Size: L 40.0 mm
- Family: **Myrmeleontidae**
- Life cycle: **One generation produced in one or more years**
- Range: **British Columbia and throughout United States**
- Food: **Adults feed on nectar and pollen, or prey on insects; larvae prey on insects**

Antlions
Myrmeleon species

Antlions are so named because of the predatory feeding habits of their pit-making larvae.

IDENTIFICATION The damselfly-like adults are brownish or grayish with darker markings and have conspicuous **antennae** that are short and thick. The long and intricately veined wings are folded rooflike over the body when at rest on vegetation during the day.

HABITS Adults are active at dusk and fly to lights in spring and summer. Wedge-shaped larvae use their **abdomens** to plow backward in the sand, leaving meandering trails, or doodles, in their wake. They find a suitable site and then move in ever tighter spirals, using their flat, shovellike head to construct a conical pit. Pits are commonly found in sandy soils under branches, rock faces, and overhangs. As an ant or other small insect approaches, the larva, or doodlebug, flips sand out

Larva

from under the feet of its hapless victim. Eventually the insect slips down the pit into the doodlebug's tonglike jaws. These jaws are first used as syringes to inject digestive enzymes into their prey and then as straws to suck out the liquefied tissues and organs.

REPRODUCTION Females drop their eggs from branches and overhangs.

SIMILAR SPECIES Only *Myrmeleon* larvae make pit traps. Other antlion larvae hunt on branches or in leaf litter. Giant antlions *(Vella)* occur across the southern half of the United States. *Glenurus gratus*, a widespread eastern species, has black and pink wing tips. Owlflies (Ascalaphidae) resemble antlions but have long, clubbed **antennae.**

> *Of Interest* Antlions make a brief appearance in Mark Twain's *The Adventures of Tom Sawyer*, when Tom finds a pit and calls out, "Doodle-bug, doodle-bug, tell me what I want to know!"

- Size: **L 15.0–20.0 mm**
- Family: **Mantispidae**
- Life cycle: **Two or three generations produced annually**
- Range: **Eastern North America**
- Food: **Adults eat small insects**

Green Mantisfly
Zeugomantispa minuta

Green mantisflies are attracted to lights and superficially resemble a cross between a praying mantis and a lacewing with a yellow stripe down its back.

IDENTIFICATION The soft-bodied adult is pale green and has threadlike **antennae**, a long **prothorax**, and a cylindrical **abdomen.** It folds its clear, green-veined wings rooflike over the body when at rest. Mature larvae spin a white to pale yellow or green cocoon that is nearly spherical or egg shaped, positioning it inside a spider egg sac, where they pupate. Some larvae spend the whole winter in spider egg sacs. Although several larvae may invade a single egg sac, usually only one reaches adulthood. Adults emerge in spring and summer and live up to 160 days. Green mantisflies occur throughout much of eastern United States, southward to Nicaragua and Panama.

REPRODUCTION Courting males and females briefly face one another before copulating, sometimes sparring with their **raptorial** forelegs. In captivity, it is not uncommon for either the male or female to cannibalize its mate. With just a single mating, females may produce up to 60 clutches of eggs, with 50–60 eggs per clutch. Eggs are laid near egg sacs of various families of spiders. The active, leggy larvae hatch in about 10 days and feed on spider eggs.

SIMILAR SPECIES *Zeugomantispa minuta* is the only green mantisfly found in North America. The brown and yellow *Climaciella brunnea* resembles a paper wasp and is widespread in North America. The genus *Plega* occurs in the West and contains four somewhat grayish species.

Of Interest Evolutionarily speaking, mantisflies are more closely related to bees, beetles, and butterflies than they are to actual mantises.

Close-up of raptorial forelegs

Beetles

Order Coleoptera

Beetles, the largest order of insects, are found in most terrestrial and freshwater habitats. Adults have chewing mouthparts and uniquely modified forewings, or **elytra.** Development is by **holometaboly.**

- Size: L 23.0–36.0 mm
- Family: Carabidae
- Life cycle: **One generation produced annually**
- Range: **Across southern Canada and most of United States**
- Food: **Adults and larvae prey on caterpillars, soft-bodied insect larvae, and earthworms**

Fiery Searcher

Calosoma scrutator

The fiery searcher, or caterpillar hunter, is one of our largest and most handsome native ground beetles.

IDENTIFICATION The **pronotum** is dark blue or violet with broad purplish, coppery, or golden-green margins. The shiny green **elytra** are purplish along the outer edges and sculpted on the surface with distinct rows of **punctures,** while the reddish-brown legs have faint metallic green or bluish highlights.

HABITS Fiery searchers overwinter as adults and emerge in spring, remaining active all summer. Beetles hide under loose bark, rocks, boards, and other debris by day, emerging at night to

search for cankerworms, tent **caterpillars,** and armyworms. They will climb trees and shrubs in search of prey and are sometimes attracted to lights. When alarmed, fiery searchers emit a pungent and burning defensive fluid from their anus. Adults live two or more years. Ground-dwelling larvae, also predatory, help control the larvae of gypsy and tent caterpillar moths.

SIMILAR SPECIES The fiery searcher resembles the forest caterpillar hunter, *C. sycophanta,* a European species introduced in the northeastern United States to control gypsy moth caterpillars. It is active in late June and July and has brilliant metallic green elytra that are narrowly reddish along the outer edges and dark brown or black hind legs. The native Wilcox's caterpillar hunter, *C. wilcoxi,* has dark metallic or bronze-green elytra with purple or golden margins.

Larva

- Size: L 6.0–10.0 mm
- Family: Scarabaeidae
- Life cycle: One generation produced annually
- Range: Across southern Canada and most of United States
- Food: Adults eat leaves; larvae eat roots

Silky Chafers
Serica species

*These are **nocturnal** beetles, and they are often found mating and feeding on the leaves and blossoms of various species of trees and shrubs, especially oaks and roses.*

IDENTIFICATION Silky chafers are oblong, convex beetles, either yellowish brown, reddish brown, or black, with nine **antennomeres** and shallowly grooved **elytra** that lack a narrow membrane bordering their tips. Under strong light, the elytra sometimes show a trace of iridescence.

HABITS Emerging in late spring or summer, these voracious beetles may congregate in large numbers and if so will defoliate fruit trees. They are commonly encountered at lights on warm evenings.

REPRODUCTION The larvae are known only for a few species of silky chafers, and they feed on the roots of grasses and other plants, apparently pruning them, eating root offshoots, rather than completely severing the taproots near ground level.

SIMILAR SPECIES There are more than 100 species of North American *Serica*, and all are superficially similar to one another. In fact, they are difficult to identify one from another without careful microscopic examination of the male reproductive organs. Ongoing studies will likely result in the splitting of North American *Serica* into several other genera.

There are two **adventive** beetle species from Asia now found in North America that resemble species of *Serica*. The Asiatic garden beetle (*Maladera castanea*) is a robust, broadly oval, chestnut-brown beetle with a dull, slight iridescent surface. Its **antennae** have ten antennomeres, and its root-feeding larvae are pests of vegetable gardens, turf, and flower beds.

The other, the pale yellowish-brown *Nipponoserica peregrina*, has a distinct dark brown band between the eyes. Like *Serica*, it also has nine antennomeres, but it has a narrow membranous border along the tips of the elytra.

- Size: L 9.0–24.0 mm
- Family: **Scarabaeidae**
- Life cycle: **One generation produced every one to four years**
- Range: **Across southern Canada and mostly central and eastern United States**
- Food: **Adults eat leaves; larvae eat roots**

June Beetles
Phyllophaga species

Also known as May beetles or (inaccurately) June bugs, June beetles are robust brown or black insects, sometimes sparsely scaly, hairy, or lightly coated with pale wax.

IDENTIFICATION The **antennae** have 9 or 10 **antennomeres,** the last three of which are flattened and fanlike and form a compact club. Underneath, the **thorax** is hairy. Adults are a common sight at lights in the Midwest, eastern, and southern United States but are only sporadically encountered in the West.

HABITS Approximately 200 species of *Phyllophaga* occur in the United States, and accurate species identification requires careful microscopic examination of the reproductive organs. All are **nocturnal** leaf feeders. Large numbers of these beetles are capable of stripping herbaceous plants and deciduous trees of their leaves. The larvae, or white grubs, eat the roots of grasses. Once of

significant economic importance in pastures and plantings of cereal crops, white grubs cause only minimal damage today.

REPRODUCTION Southern populations of *Phyllophaga* take as little as a year to complete their development, while those farther north may require four years to reach adulthood. Beetles emerge from their pupae in late summer or fall but remain in their pupal chambers until the following spring or early summer.

SIMILAR SPECIES Species of *Phyllophaga* are most likely to be confused with much smaller chafer species in the genus *Diplotaxis,* but the underside of their thorax is not hairy. Masked chafers (*Cyclocephala*) have **mandibles** visible from above but hidden from view in *Phyllophaga.*

Of Interest Planting deep-rooted legumes like alfalfa and clover instead of ornamental grasses can reduce the number of June beetle eggs laid in a lawn or garden.

- Size: L 18.0–31.0 mm
- Family: Scarabaeidae
- Life cycle: One generation produced every two to five years
- Range: Western North America to the Dakotas, Nebraska, and Kansas
- Food: Adults eat conifer needles; larvae eat roots

Ten-Lined June Beetle
Polyphylla decemlineata

These beetles are most apparent as they fly at dusk and in the evening. They are strongly attracted to lights, especially the males.

IDENTIFICATION The ten-lined June beetle is reddish brown to black with distinctive, smooth-edged stripes of variable widths on the **elytra** that are composed of cream to brownish-yellow **scales.** The male's antennal club is composed of seven long and curved **antennomeres,** while that of the female is composed of five relatively short and straight antennomeres. Males are capable of spreading their **antennae** out, fanlike.

HABITS Adults emerge in late spring and are active throughout the summer. When disturbed, they make a loud squeaking noise by rubbing the rough edges of their elytra against special plates on the **abdomen.** Ten-lined June beetles

are the most common widely distributed *Polyphylla* in western North America and occur in coastal plain, chaparral, grasslands, desert scrub, oak-juniper woodland, mixed woodland, and coniferous forest habitats.

REPRODUCTION The life cycle may take two to five years, depending on local conditions. The subterranean C-shaped grubs consume roots of both wild and cultivated plants and are sometimes pests in agricultural crops, orchards, and pine plantations. Mature larvae pupate deep in the soil to avoid freezing and move back to the surface in spring to feed and complete their development. The pupal stage lasts up to five weeks.

SIMILAR SPECIES *Polyphylla* are sometimes difficult to identify by species, especially old or worn beetles. The ten-lined June beetle is distinguished from the other 31 species of *Polyphylla* that occur in North America by the lack of erect **setae** on the **pronotum,** the distinct and smooth-edged elytral stripes, and western distribution.

Male with spread antennae

- Size: L 8.9–11.8 mm
- Family: Scarabaeidae
- Life cycle: One generation produced annually
- Range: Maine to Georgia, west to South Dakota and Texas
- Food: Adults consume leaves, flowers, and fruits

Japanese Beetle
Popillia japonica

The Japanese beetle, an **adventive** *species from Japan, is a highly destructive pest of ornamental and crop plants.*

IDENTIFICATION Dark metallic green with reddish-brown **elytra,** it is the only beetle in North America with a pair of distinctive spots on the **pygidium** and five white tufts along each side of the **abdomen.**

HABITS The **diurnal** adults feed on more than 300 species of plants in gardens, pastures, vacant lots, and field crops. They eat flowers and the soft tissues of the upper surfaces of leaves, in between the veins.

REPRODUCTION Eggs are deposited in the soil. The grubs hatch in about two weeks and begin feeding on roots, becoming serious pests of turf in lawns, parks, and golf courses. They will also attack the roots of ornamental plants and vegetables. The third **instar** larvae

overwinter deep in the soil and resume feeding near the surface in spring. The pupal stage lasts about 8–20 days, and the beetles start emerging in late spring.

SIMILAR SPECIES Japanese beetles are occasionally confused with two other scarab beetles, the diurnal *Strigoderma arbicola* and the **nocturnal** *Callistethus marginatus*. Neither of these two indigenous North American species has white abdominal markings, however.

Of Interest Since their discovery in New Jersey in 1916, Japanese beetles have spread throughout most of the eastern United States and into states along the western banks of the Mississippi River; localized infestations occasionally occur farther west. The USDA operates a Japanese Beetle Quarantine to protect agriculture in the West and examines aircraft to reduce the spread of the pest westward.

Swarms of adult on rose

- Size: L 18.0–27.0 mm
- Family: Scarabaeidae
- Life cycle: One generation produced every two years
- Range: Ontario and Maine to Florida, west to South Dakota and Texas
- Food: Adults eat wild and cultivated grape leaves; larvae eat rotting wood

Grapevine Beetle
Pelidnota punctata

Starting in early summer, the adults of this species can be found clinging to the stems and the undersides of leaves of grapevines, feeding on leaves and fruit.

IDENTIFICATION The grapevine beetle, also known as the spotted grape beetle, is easily distinguished from other large scarab beetles by its uniformly yellowish to reddish-brown **pronotum** and **elytra.** The elytra may or may not be flanked by black spots that are variable in size. The black legs have greenish reflections. Northern populations tend to have larger, more distinct spots, while those in the South, especially in Florida, sometimes lack spots altogether. The underside ranges in color from yellowish brown to black.

HABITS These beetles begin flying at dusk to search for food and mates. They are commonly attracted to lights. Adults feed on the leaves and fruit of grapes, both cultivated and wild, but rarely cause significant crop damage.

REPRODUCTION After mating, the females lay their eggs in moist soil at the base of hardwood stumps or under logs, including apple, elm, oak, sycamore, and walnut. The subterranean larvae develop in rotting stumps or in hollowed-out tunnels chewed inside lateral roots just beneath the soil surface. Mature grubs construct pupal cells from surrounding wood fragments, grass, and other bits of nearby vegetation. The entire life cycle takes two years; the adults live for about a month.

SIMILAR SPECIES Most *Pelidnota* species occur in the New World tropics. The only other North American species is *P. lugubris,* a much smaller and completely black beetle found in southeastern Arizona and southwestern New Mexico. The genus *Cotalpa,* which includes several species found in the eastern United States and the Southwest, are similar in size to *P. punctata* but are yellowish or creamy yellow in color.

Larva and pupa

- Size: L 11.0–14.0 mm
- Family: Scarabaeidae
- Life cycle: One generation produced annually
- Range: New York to Florida, west to Iowa, Kansas, and Texas.
- Food: Adults probably eat leaves; larvae eat roots

Northern Masked Chafer
Cyclocephala borealis

Northern masked chafers are common throughout eastern North America, where both males and females are frequently encountered at lights during the night.

IDENTIFICATION These shiny beetles, oblong-oval in shape, are mostly dull yellowish brown in color with darker-colored markings between the eyes. The undersides of both the **thorax** and the **elytra** are sparsely covered with scattered **setae.**

HABITS Dozens to hundreds of these beetles will buzz low over grassy areas in yards, parks, and golf courses in search of mates on warm summer nights starting in mid-June.

REPRODUCTION Small knots of amorous males grapple with one another on the ground, striving to mate with a single female in response to her releasing an attractive sex **pheromone.** The sole successful male secures her by grasping the edges of her **elytra** with his enlarged front claws. Soon after mating, the female digs down into the turf in order to lay her eggs. The subterranean grubs, shaped like a "C," hatch in about two weeks and begin feeding on the roots of grasses. Mature larvae dig deep into the soil in the fall and hibernate over the winter. In spring, they move closer to the surface to feed before pupating in late spring. The grubs may become serious pests of turf and cereal crops, especially in late summer and fall and again in the spring.

SIMILAR SPECIES Fourteen species of *Cyclocephala* occur in Canada and the United States. In eastern North America, *C. borealis* is distinguished from *C. lurida* by the sparse setae on the elytra. A number of *Cyclocephala* species are common in gardens, parks, and golf courses in western North America, including *C. hirta, C. longula,* and *C. pasadenae.*

Larvae under turf

- Size: L 10.0–17.0 mm
- Family: Scarabaeidae
- Life cycle: One generation produced annually
- Range: Across southern Canada and United States
- Food: Adults and larvae eat roots

Carrot Beetle
Tomarus gibbosus

Carrot beetles are sometimes attracted in large numbers to lights at night in the spring and throughout the summer—even in late winter in the West—and are reported to live in the soil year-round.

IDENTIFICATION A common native species, the carrot beetle is robust, oval, and moderately shiny reddish brown to black, with a narrowed **clypeus** that ends in two small upturned teeth. The **mandibles** are each flanked with three small toothlike projections. The front of the **pronotum** has a small **tubercle** and dimple just behind the front margin. The **elytra** each have 7–10 rows of pitlike **punctures.** The front **tarsi** of the males are slender, while the hind tarsi in both sexes are nearly equal in length to the **tibia.**

HABITS Both adults and larvae are found in sandy soils rich in plant detritus, where they feed on the roots of many different kinds of plants. The C-shaped larvae not only feed on the roots of pigweed, cereals, and sunflowers, but they also consume decaying plant materials. They are commonly found under old cattle droppings.

SIMILAR SPECIES You have to look closely to see the differences between this species and other scarab beetles. In the East, the characteristics of the head, pronotum, and legs of *T. gibbosus* will serve to distinguish it from other related scarab beetles. In the Southwest, the carrot beetle's narrowed clypeus ending in two small upturned teeth will separate it from the similar reddish-brown *Oxygrylius ruginasus* with only a single upturned tooth.

Of Interest Feeding activities of the carrot beetle are reported to cause injury to carrots, celery, corn, cotton, elm, oak, parsnip, potatoes, sugar beets, and several other crops.

Larva

- Size: L 40.0–60.0 mm
- Family: Scarabaeidae
- Life cycle: One generation produced every two years
- Range: New York to Florida, west to southern Illinois, western Arkansas, and eastern Texas
- Food: Adults drink sap infused with microorganisms and eat fruit; larvae eat rotting wood

Eastern Hercules Beetle

Dynastes tityus

*The male of this impressive species has one curved horn on the head and three horns on the **pronotum**, the largest of which projects forward over the head.*

IDENTIFICATION Also called a rhinocerus or unicorn beetle, this large, olive-green species is usually mottled with irregular black spots on its smooth and shiny elytra. The females have similarly markings, but they have no horns.

HABITS Adults emerge early in the summer and are sometimes attracted to lights. They will occasionally aggregate on ash branches, especially near wounds where sap is leaking, its fragrance attracting females. There the males congregate and use their horns to battle rival males.

REPRODUCTION Males and females frequent tree holes that serve as egg-laying sites. The larvae do not damage living trees, preferring instead to feed on the rotten and crumbling **heartwood** of oak, cherry, black locust, and willow; they will occasionally use pine. Pupation takes place in late summer inside a cell constructed from the larvae's **feces** mixed with plant debris. Adulthood is reached in several weeks, but they remain in the cell until the following summer, when they emerge as adult beetles.

SIMILAR SPECIES Other species of large horned beetles include the western hercules beetle, *D. grantii*, which occurs in southwestern Utah and Arizona, east to New Mexico. Adults feed on the sap of velvet ash branches during the day. The dark, reddish-brown males of *Strategus*, commonly known as ox beetles, usually have three distinct horns on the pronotum; they occur from southeastern Arizona to the eastern United States. Male rhinoceros beetles—the eastern (*Xyloryctes jamaicensis*) and southwestern (*X. thestalus*)—are dark brown and so named because each bears a single horn on the head.

Adult female

- Size: L 20.0–30.0 mm
- Family: Scarabaeidae
- Life cycle: **One generation produced annually**
- Range: **Central and southern California east to Colorado and western Texas**
- Food: **Adults eat ripened fruit and plant sap; larvae eat decaying vegetation**

Green Fig Beetle
Cotinis mutabilis

The green fig (or peach) beetle abounds near orchards or in neighborhoods with backyard fruit trees and compost piles.

IDENTIFICATION These velvety green beetles are variably marked with orangish brown or tan on the **elytra,** especially on the outer margins, while the underside is shiny iridescent green. Both males and females possess a short, broad horn that they use to pierce the skin of soft fruits.

HABITS They fly noisily through backyards or meander over turf and accumulations of decaying plant debris in yards, orchards, and parks. Singly or in small groups, they will attack apricots, peaches, figs, nectarines, and other soft fruits either damaged by birds or fallen to the ground.

REPRODUCTION Females lay their eggs in compost heaps and under mulch, haystacks, and manure heaps. In spite of their well-developed legs, the C-shaped grubs crawl outstretched on their backs, resulting in their nickname "back crawlers." They use rows of stiff **setae** across their backs to gain purchase. The first two larval stages are completed by autumn, and the third **instar** continues feeding in the spring. Pupation takes place within a protective case constructed from the grub's **feces.** Adults emerge in June, reach their peak activity in midsummer, and remain active until November.

SIMILAR SPECIES Green fig beetles have the tip of their abdomen and hind legs uniformly dark. In another species that is similar in form and habit, the green June beetle *(C. nitida),* these features are bicolored.

Of Interest Green fig beetles can be clumsy fliers and find themselves caught in swimming pools or birdbaths. As long as they get out of the water quickly they can dry off and continue flying.

Larva on its back

- Size: L 24.0–44.0 mm
- Family: Elateridae
- Life cycle: Probably one generation produced annually
- Range: Eastern North America
- Food: Adults and larvae prey on larvae of wood-boring beetles

Eyed Click Beetle
Alaus oculatus

Lying on their backs, these insects will right themselves by flipping right side up with a click, and hence their name.

IDENTIFICATION The upper surface of adult eyed click beetles is covered with black-and-white or pale yellowish-white **scales.** The pair of large scaly eyelike spots on the **pronotum,** each narrowly ringed with white scales, are not special organs of any kind. They were once thought to deter would-be **predators** by making them think they had lost the element of surprise. The mature yellowish-brown larvae have tough **exoskeletons** and reach 2 inches (50 mm) in length.

HABITS Both the adults and the larvae are predaceous, and they can be found year-round, especially in spring and summer, on trees or under the loose bark of decaying logs and stumps infested with wood-boring beetle larvae. Adults are more likely to be found on the trunks and branches of oak, beech, willow, sycamore, tulip, and other deciduous trees infested with wood-boring beetles in the summer.

SIMILAR SPECIES *Alaus myops,* the only other widespread eyed click beetle in the East, is associated with pines and distinguished from *A. oculatus* by the dominance of grayish or slightly brownish scales and small elliptical eyespots narrowly ringed with gray scales. *Alaus patricius* has greatly reduced eyespots and **elytra,** each tipped with a pair of spines; its range is restricted to the Florida Keys. Another species, *A. melanops,* is widespread in coniferous forests in the West. *A. lusciosus* and *A. zunianus,* both found in the Southwest, are mostly black with large, bold, white markings.

Of Interest These beetles are also sometimes attracted to surfaces that have been freshly painted, perhaps because the solvents in the paint resemble the species' sex **pheromones.**

Larva

- Size: L 12.0–20.0 mm
- Family: **Cantharidae**
- Life cycle: **Probably one generation produced annually**
- Range: **British Columbia to California**
- Food: **adults prey on sap-feeding insects; larvae prey on insects**

Brown Leatherwing Beetle
Pacificanthia consors

*Often encountered at porch lights in late spring and early summer, **nocturnal** brown leatherwings prey on sap-feeding insects.*

IDENTIFICATION This species of soldier beetle is reddish brown with soft, leathery, gray **elytra** and partly dark legs. They superficially resemble fireflies, with their broad heads clearly visible in front of the **pronotum** when viewed from above.

HABITS Very little is known about the life history of these beetles. The nocturnal adults prey on the citrus mealybug (*Pseudococcis citri*) and other insects. They are attracted to porch lights, especially in neighborhoods near coastal chaparral and oak woodland habitats. These beetles spend their days hiding under loose bark or in leaf litter, and, when roughly handled or crushed, they give off a strong unpleasant odor.

REPRODUCTION The larvae live and develop in leaf litter, where they probably hunt for small insects.

SIMILAR SPECIES The large size, broad head, and body and leg colors will distinguish brown leatherwing beetles from most other soldier beetles in the West. The habits of most soldier beetles are poorly known, although some species are insect **predators.** The short-lived adults of other species are found on flowers and foliage during the day or at lights at night.

Of Interest Soldier beetles such as this are sometimes found dead with their **mandibles** embedded in stems or leaves. They have likely been infected by fungal pathogens in the plants they were eating. Characterized by open wings and contorted bodies, these apparently violent deaths likely enhance dispersal of the killer fungus spores.

Larva

- Size: L 9.0–12.0 mm
- Family: Cantharidae
- Life cycle: One generation produced annually
- Range: Maritime Provinces to Florida, west to Ontario, Colorado, and Texas
- Food: Adults eat pollen; larvae prey on insects

Goldenrod Soldier Beetle
Chauliognathus pensylvanicus

Individuals or clusters of goldenrod soldier beetles feeding and mating on flowers in late summer and fall are familiar sights in eastern North America.

IDENTIFICATION The head of these elongate and soft-bodied beetles is entirely black, while the **pronotum,** wider than long, has only a broad black patch. Each **elytron** has a variable black spot on the tip that may extend all the way to its base. The specific name is Latin for "of Pennsylvania." Swedish entomologist Charles De Geer scientifically described this beetle in 1774, when the colony was commonly spelled "Pensylvanie," hence the spelling of this species' scientific name.

HABITS Adults feed on pollen from various flowers, especially goldenrod (*Solidago*), growing in gardens, parks, fields, meadows, and along roadsides and woodland edges in late summer and early fall.

REPRODUCTION Females deposit their eggs in leaf litter, where the larvae develop and probably prey on small insects and their eggs.

SIMILAR SPECIES The broad pronotum and late summer-to-fall activity period of the goldenrod soldier beetle distinguish it from another common and similarly colored beetle in the region, the margined leatherwing, *C. marginatus.* This species, active in the spring and early summer, has an orange head with a black V-shaped mark and a pronotum that is longer than wide.

Of Interest The goldenrod soldier beetle is a favorite for entomologists conducting laboratory experiments. It has been used as a subject in studies of mating behavior, variation in color pattern, dispersal, and insect genetics.

Mating pair

- Size: L 9.0–15.0 mm
- Family: Lampyridae
- Life cycle: One generation produced annually
- Range: New York to northern Florida, west to South Dakota and Texas
- Food: Adults do not feed; larvae prey on invertebrates

Big Dipper Firefly
Photinus pyralis

The big dipper firefly is the best known and most widely distributed firefly in eastern North America. Called fireflies or lightning bugs, they are neither flies nor true bugs but are soft-bodied beetles.

IDENTIFICATION Their heads are completely covered by a hoodlike **pronotum** when viewed from above. The middle of the big dipper firefly's pronotum is pinkish with a variable black spot inside, while the soft dark brown or blackish **elytra** have pale margins. The pale light-producing organs are clearly visible on the underside of the **abdomen.** Those of the female take up part of one segment, while those of the male consist of two entire segments.

HABITS Male big dipper fireflies emerge just before sunset and fly in a series of low vertical loops along the edges of gardens and parks. They direct their distinctive J-shaped flashes of greenish-yellow light at females perched low in the nearby vegetation. Receptive females respond to flashing males with their own signals of light.

SIMILAR SPECIES Another species widespread in eastern North America is the Pennsylvania firefly, *Photuris pennsylvanicus,* recognized by the combination of a dark anchor-shaped pattern flanked by two reddish spots on the pronotum and an oblique pale stripe that partially or almost entirely covers the length of each elytron.

> *Of Interest* Fireflies represent one of nature's best known examples of bioluminescence. Each species of firefly has its own flash patterns, thus reducing the chances of mating with others. Big dipper fireflies display a J-shaped flash pattern, while Pennsylvania fireflies use a "dot-dash" pattern.

Light-producing organ

LIGHTING UP THE NIGHT

At the height of summer in eastern North America, amorous male fireflies, or lightning bugs, take to the warm night air to emit precisely timed bursts of species-specific yellowish, greenish, or orange light. Receptive females, often perched low among the nearby shrubs, respond to these Morse code–like signals, each species with its own distinctive flash pattern.

FIREFLY CHEMISTRY

Glowing fireflies dot the woods in Great Smoky Mountains National Park in Tennessee. This spectacular mating display usually occurs between May and June.

The production and emission of light in fireflies and other living organisms by chemical reaction is called bioluminescence. The **adenosine triphosphate** (ATP) that powers living cells in all organisms is a key ingredient for bioluminescence in fireflies. The enzyme **luciferase** serves as a binding agent and attaches the chemical pigment **luciferin** to ATP. Oxygen (O_2) supplied via the **tracheoles** is temporarily diverted from the cell's mitochondria to its light-producing organs in the

A big dipper firefly, the most common species in North America, rests on a soybean plant, the glow of its abdomen illuminating the leaf.

presence of nitric oxide. Oxygen binds with a carbon atom of the newly energized luciferin, temporarily kicking an electron into a higher orbit and releasing both carbon and oxygen as carbon dioxide (CO_2). As the electron slips back into its original orbit, energy is released as a tiny flash of light.

This chemical reaction, repeated in thousands of light-producing cells located in special organs in the firefly's **abdomen,** produces enough light to be clearly visible to the human eye. Controlled ultimately by the insect's nervous system, the amount of oxygen reaching the light-producing organs determines the brightness and duration of the signal.

USEFUL BEYOND SEX

Since the non-reproductive stages (eggs, larvae, pupae) of fireflies also glow, bioluminescence is more than just about sex. All life stages in fireflies contain lucibufagins, bitter steroids that can make birds and mammals ill. Once sickened, these **predators** quickly learn not to make a meal of glowing insects. Thus bioluminescence also serves an **aposematic** function in fireflies by warning potential predators of their distastefulness.

ENERGY EFFICIENCY

The "cold light" of fireflies is produced by an incredibly efficient system. All but 2 percent of the energy that goes into the system is released as light, with only a fraction lost as heat, compared with 23 percent for compact fluorescent lightbulbs and up to 95 percent for incandescent bulbs.

LEARNING FROM FIREFLIES

Luciferase is used in medical research to identify blood clots and monitor specific cells implicated in heart disease, cancer, and other potentially life-threatening conditions. Bioluminescence in fireflies has also inspired the development of more efficient lighting systems, including an improved LED bulb that emits 55 percent more light.

- Size: L 4.8–7.5 mm
- Family: Coccinellidae
- Life cycle: One generation produced annually
- Range: North America
- Food: Adults prey on aphids and related sap-feeding insects; larvae prey on insects

Multicolored Asian Lady Beetle

Harmonia axyridis

The multicolored Asian lady beetle is a species from Asia that sports dozens of combinations of color and pattern.

IDENTIFICATION The **pronotum** is mostly dull white and marked with five black spots that often merge to form a black "M" or a solid trapezoid. The pale yellow-orange to bright–red orange **elytra** may be spotless or have as many as 22 black spots on fully marked individuals.

HABITS Native to Japan, Korea, and eastern Russia, this species was introduced into the United States beginning in 1916, but established populations were not found until 1988 in Louisiana. Adults are found year-round on trees and shrubs and are sometimes attracted to light. When disturbed or crushed, they emit a strong-smelling yellowish fluid from their leg joints (**reflex**

bleeding) to discourage **predators;** it may stain clothing and walls. Considered beneficial, these beetles may become a nuisance when they gather on or enter homes and outbuildings in late fall and winter to escape freezing temperatures.

REPRODUCTION Eggs laid in spring hatch in three to five days. The larvae prey on aphids and other soft-bodied insects. Mature larvae are gray to black with spiny, cone-shaped projections and an orange stripe down each side of the body.

SIMILAR SPECIES Multicolored Asian lady beetles are sometimes difficult to distinguish from other species of lady beetles because of the variability in their colors and patterns. The simplest method is to look for the "M" (or "W") pattern.

Of Interest Recent years have seen a population boom, thought to be from a new source of beetles accidentally introduced from an Asian freight ship.

Various color morphs

- Size: L 4.0–7.0 mm
- Family: **Coccinellidae**
- Life cycle: **One generation produced annually**
- Range: **North America**
- Food: **Adults eat aphids, also pollen and other plant materials; larvae prey on insects**

Convergent Lady Beetle
Hippodamia convergens

This is North America's most familiar "ladybug" (an entomological misnomer, since this is a beetle, not a bug).

IDENTIFICATION This beetle's body is somewhat oval and convex, with **femora** clearly sticking out beyond the margins of the orange **elytra.** The black **pronotum** has convergent pale markings, with white borders of more or less even width along the sides. Each elytron usually has six small and distinct spots present, but these markings may be reduced in number or absent altogether.

HABITS Adults become active in spring and summer and are found year-round in a wide variety of habitats on vegetation. In the West, adults escape hot, dry conditions by flying up cooler river canyons into the mountains in search of food. They eventually overwinter in large aggregations and return to their valley feeding grounds in spring. Adults are collected at overwintering sites in California and sold to control aphids and other pests, but their larvae are more effective **predators.**

REPRODUCTION Females lay batches of 15–30 bright-yellow eggs on plants in spring and summer. The long, flat larvae are velvety black or gray with small orange spots on the first and fourth abdominal segments. These aphid predators **molt** four times in about a month. The orange pupae with black spots are frequently attached to plants, walls, and fences. The pupal stage lasts about a week.

SIMILAR SPECIES The convergent lady beetle is the most common and abundant species of *Hippodamia* and is distinguished from any others by its distinct pale markings on the pronotum.

Of Interest The convergent lady beetle is the official state insect of North Dakota.

Larva

- **Size:** L 6.4–8.1 mm
- **Family:** Coccinellidae
- **Life cycle:** One to three generations produced annually
- **Range:** Eastern North America
- **Food:** Adults and larvae eat bean plant leaves

Mexican Bean Beetle
Epilachna varivestis

Mexican bean beetles are plant-feeding lady beetles; both adults and larvae are garden and crop pests, especially during late summer.

IDENTIFICATION The oval and somewhat convex adults are pale brownish yellow to dark reddish yellow. The **pronotum** is spotless, while the **elytra** each have eight round spots arranged in three rows, none of which reach the **elytral suture.**

HABITS Originally from the southeastern United States and Mexico, Mexican bean beetles have gradually expanded their range and are now established throughout most areas east of the Rockies. Adults overwinter in protected areas along garden and field edges and emerge the following spring to feed and mate.

REPRODUCTION Females lay masses of yellow eggs on the undersurfaces of leaves. The spiny yellow or orange-yellow larvae skeletonize the lower surface of the leaf, where they will pupate in about three weeks. Their feeding activities primarily damage the leaves of garden beans and field beans, but they will also occasionally attack soybeans and cowpeas. As many as three generations are produced annually in the Southeast, where they are most abundant.

SIMILAR SPECIES The squash beetle, *E. borealis*, is similar in form and habit to the Mexican bean beetle. It is distinguished by having four spots across the pronotum and eight large oval spots on the elytra, two of which merge together on the elytral suture. Also in the eastern United States, especially the Atlantic states, the adults and larvae of this species are only minor pests of squash plants.

Of Interest Interspersing flowers or potato plants with bean plants can help prevent Mexican bean beetles from eating away an entire crop of beans in the garden.

Larva

SEARCHING FOR C-9

Lady beetles, known to many as ladybugs, are beloved creatures considered harbingers of good luck. Gardeners and farmers have long encouraged them to take up residence on their lands to benefit from their services as **predators** of aphids and other insect pests. But, sadly, several native lady beetles are on the decline.

DECLINE OF THE NINE-SPOT

The nine-spotted lady beetle, or C-9 (*Coccinella novempunctata*), is one of the species whose population numbers have declined sharply over the years. Designated by the state of New York as its official insect in 1989 because of its seeming abundance and importance to agriculture, this species has since disappeared from the state. Surveys show that this once widespread species now occupies a small fraction of its former range across southern Canada and much of the United States. Why?

ALIEN INVASION?

Nearly 200 species of lady beetles have been intentionally introduced into North America to control insect pests on farms and in orchards. Of those 200, about 30 have become established, including the very common multicolored Asian (*Harmonia axyridis*) and seven-spotted (*Coccinella septempunctata*) lady beetles. Are C-9 and other native species unable to compete with these introduced species for food? Might they be susceptible to diseases harbored by non-native species? Has the steady loss of farmland in the region contributed to their disappearance? Their rapid decline of since the 1980s is still a mystery.

Nine-spotted lady beetles were once abundant across North America, but since the introduction of non-native lady beetles and habitat loss in their range, their numbers have fallen.

LOST LADYBUG PROJECT

After 14 years without a trace in the northeastern United States, C-9 was rediscovered by two children in northern Virginia in 2006. This and additional sightings across North America inspired entomologists at Cornell University to establish a new citizen science program, the Lost Ladybug Project (*lostlady bug.org*), to teach the importance of protecting biodiversity and recruit members of the public to document the ongoing status of C-9 and other native lady beetles in decline.

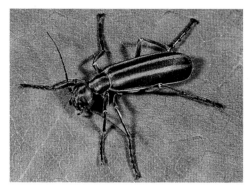

- Size: L 9.0–18.0 mm
- Family: Meloidae
- Life cycle: One generation produced annually
- Range: Eastern North America
- Food: Adults eat leaves, especially of plants in nightshade family; larvae eat grasshopper eggs

Striped Blister Beetle

Epicauta vittata

These beetles are fond of plants in the nightshade and pea families. They will attack tomato, potato, and bell pepper plants as well as low-growing weeds.

IDENTIFICATION Striped blister beetles are brownish yellow with bold, black stripes. Both the antlike head and **pronotum** bear a pair of dark markings, while the **elytra** each have two (northeastern populations) or three (western, southeastern populations) black stripes. Beetles in the Northeast tend to be darker with two narrowly separated stripes on each elytron. In the Southeast, especially on the coastal plain, they tend to have more distinctly separated middle and outer stripes, hence three stripes.

HABITS Adults are found in gardens, parks, and fields during summer and are sometimes attracted to lights at night

in large numbers. Large mating swarms of beetles cause considerable economic damage when they consume the fruits, flowers, and leaves of vegetable and forage crops.

REPRODUCTION Mating pairs remain coupled for several hours. The larvae develop by **hypermetamorphosis.** The leggy first **instar** larva, or **triungulin,** actively seeks grasshopper egg pods in the soil, then transforms into a sedentary grub that consumes the grasshopper eggs.

SIMILAR SPECIES The striped blister beetle resembles three other striped species of American *Epicauta*: *E. occidentalis* (mostly south-central states), *E. temexa* (Texas), and *E. abadona* (southern Arizona).

Of Interest Blister beetles engage in a defensive strategy known as **reflex bleeding.** When disturbed, they secrete droplets of bright-yellow, caustic **hemolymph** from their leg joints.

Variations in elytral patterns

- Size: L 24.0–55.0 mm
- Family: **Cerambycidae**
- Life cycle: **One generation produced every three to five years**
- Range: **Western North America**
- Food: **Adults do not feed; larvae eat tree roots**

California Prionus
Prionus californicus

This large, reddish brown beetle with long antennae is found in foothill and mountain habitats.

IDENTIFICATION The **antennae** are conspicuous, especially those of males. Saw-toothed with 12 **antennomeres,** they measure more than two-thirds the body length of the male, while those of the female are relatively slender and only about half as long as the body. The sharply margined **pronotum** of both sexes has only three sharp spines on each side.

HABITS Adults begin to emerge in summer, becoming active at dusk and in the evening, and commonly fly to lights.

REPRODUCTION Large females may carry up to 1,200 eggs. When fully developed, the large root-feeding larvae, sometimes called giant root borers, may grow up to 80 mm in length, as thick as a man's finger. They tunnel along the outside of roots of living deciduous trees to feed and are especially fond of oaks, but will also attack cottonwood, madrone, and occasionally fruit trees. Mature larvae tunnel well away from the root system to pupate.

SIMILAR SPECIES In California, a similar species, *P. lecontei*, has 13 antennomeres and a plump and flightless female. Several other *Prionus* species are found across the United States. Other similarly large western longhorn beetles include species of *Derobrachus* (deserts) and *Trichocnemis* (mountains), all of which have only 11 antennomeres and four or more **pronotal** spines on each side.

Of Interest All these large beetles sometimes harbor tiny arachnids called pseudoscorpions. Resembling tailless scorpions, these mite **predators** rely on large beetles for transport to decaying and mite-infested stumps.

- **Size:** L 8.0–15.0 mm
- **Family:** Cerambycidae
- **Life cycle:** One generation produced annually
- **Range:** Eastern North America to Colorado
- **Food:** Adults eat milkweed leaves; larvae eat milkweed roots

Red Milkweed Beetle

Tetraopes tetrophthalmus

Apparently gregarious, red milkweed beetles commonly cluster in groups, feeding and mating on milkweeds in late spring and summer.

IDENTIFICATION The bright-red milkweed beetle has bold, black markings on the **pronotum** and **elytra.** Both the genus and species names mean "four eyes," referring to the two black **compound eyes** that are each completely divided in two. The middle of the pronotum has a raised hexagonal callous surrounded by four black spots.

HABITS This beetle seeks plants with multiple large flower clusters and consumes buds, blossoms, and leaves. Before feeding on a mature leaf, the beetles chew into the base of the midrib to drain out the milkweed's sap, thus limiting their exposure to the plants' sticky toxin. When handled, they **stridulate,** or produce a

squeaking sound by rubbing a collarlike scraper on the pronotum against a filelike ridge on the adjacent **mesonotum.**

REPRODUCTION Females deposit their eggs in dried milkweed stalks. The larvae usually tunnel through the soil from root to root as they feed, and they may briefly tunnel inside milkweed roots. Mature larvae pupate near the soil surface. Adults in captivity have been observed to live about two months.

SIMILAR SPECIES *Tetraopes tetrophthalmus* is distinguished from the other four eastern species in the genus by its red color and four distinct black spots on the elytra. Another milkweed longhorn beetle, *T. femoratus,* is widespread in western and central United States.

Of Interest Red milkweed beetles are thought to incorporate some of their host plant's toxins into their bodies, making them an unpleasant snack for any potential **predators.**

Close-up of head showing divided compound eye

- Size: L 8.5–12.0 mm
- Family: **Chrysomelidae**
- Life cycle: **One generation produced annually**
- Range: **Across southern Canada and United States**
- Food: **Adults and larvae eat leaves of morning glories and their relatives**

Argus Tortoise Beetle
Chelymorpha cassidea

Named after the hundred-eyed giant in Greek mythology, this is one of North America's largest leaf beetles, sometimes mistaken for a very large lady beetle.

IDENTIFICATION These turtle-shaped beetles are typically brick red when fully developed (**teneral** specimens are yellowish), with 4–6 black spots across the **pronotum** and a total of 13 spots on the **elytra,** including a spot across the base of the **elytral suture.**

HABITS In gardens, old fields, and other open habitats, adults emerge in spring to feed and mate and remain active until early summer. Both adults and larvae feed mainly on plants in the morning glory family and damage sweet potato crops by chewing holes in the leaves. They are sometimes mistakenly called milkweed tortoise beetles, even though they do not attack milkweeds. In fall,

after a brief period of feeding, the new generation of adults seeks dry and protected places, such as under bark or in leaf litter, to overwinter.

REPRODUCTION Females lay their slender, stalked yellow eggs singly or in batches on the undersides of leaves. The spiny, flattened larvae chew small holes in the leaves, protected by an umbrella-like shield composed of their own cast **exoskeletons** and **feces.** The pupa is attached to the leaf, festooned with feces and other debris gathered throughout the larva's life.

SIMILAR SPECIES *Chelymorpha phytophagica*, a similar species in both appearance and habit, occurs in the Southwest.

> *Of Interest* In Greek mythology, Argus Panoptes was a hundred-eyed giant who was a servant of the goddess Hera. It is said that after his death, she preserved his hundred eyes forever in the feathers of a peacock's tail.

Larva

A TALE OF TWO INVASIVES

Invasives are aggressive species of plants, animals, and other organisms that have been introduced to areas outside their native region. Free of competition, they harm native species, disrupt the economy, or negatively impact human health. Two of North America's most notorious invasive beetles were accidentally introduced from Asia.

EMERALD ASH BORER

The bright–metallic green emerald ash borer, *Agrilus planipennis*, first discovered in Detroit, Michigan, and Windsor, Ontario, during summer 2002, likely arrived in untreated wood packing materials from eastern Asia. The wood-boring larvae have killed millions of trees and threaten all species of ash across the continent. Control efforts involve bans on moving firewood and the introduction of predatory wasps imported from China.

Like the Asian longhorn beetle, the emerald ash borer lays its eggs in tree bark. Developing larvae disrupt the flow of water and nutrients, ultimately killing the tree.

ASIAN LONGHORN BEETLE

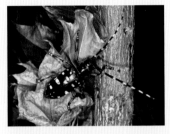

The shiny black Asian longhorn beetle, *Anoplophora glabripennis*, usually has irregular white spots, bluish or white legs, and long **antennae** ringed in pale blue or white. First reported in New York in 1996, it was likely introduced via untreated wood used to crate heavy equipment. Established in portions of the Northeast and upper Midwest, this species threatens to spread to other parts of North America. The larvae tunnel through trunks and limbs, weakening and killing forest and street trees. Eradication involves cutting down, chipping, and burning infested trees.

Asian longhorn beetles, brought to North America in the 1990s, decimate hardwoods. Infested trees are destroyed, the areas in which they are growing quarantined.

- Size: L 9.0–11.5 mm
- Family: Chrysomelidae
- Life cycle: One generation produced annually
- Range: North America
- Food: Adults and larvae eat leaves of potatoes, eggplants, and their relatives

Colorado Potato Beetle

Leptinotarsa decemlineata

A serious pest of potatoes since about 1840, this is one of the few North American insects that has become established in Europe and Asia.

IDENTIFICATION Colorado potato beetles are strongly rounded, hump-backed, and boldly marked. The orange-brown **pronotum** has a black U-shaped mark flanked by six small spots, while the yellowish **elytra** are each boldly marked with four black stripes, with the second and third stripes out from the **elytral suture** joining at the tips.

HABITS Adults and larvae eat the leaves of potatoes, eggplants, and other plants in the nightshade family in gardens, parks, and old fields. Younger larvae feed in groups, while older ones typically eat alone. Mature larvae crawl down the food plant and burrow into the soil to pupate. Overwintering adults emerge in spring to search for food and mates.

REPRODUCTION The females attach masses of 20–60 eggs on the undersides of leaves. The plump, humpbacked larvae are orange red, pinkish, or reddish and have a double row of black spots along each side.

SIMILAR SPECIES The Colorado potato beetle strongly resembles another eastern species, the false potato beetle, *L. juncta*, in both appearance and habit. In *L. juncta*, the two outer-most elytral stripes (rather than the two middle stripes) are joined at the tips. These adults and larvae also eat leaves of nightshades, including horse nettles and ground cherry. Their plump white larvae have a single row of black spots on each side.

Of Interest These beetles have the ability to develop resistance to chemical pesticides, which contributes to their status as a pest around the world.

Larva

- Size: L 4.6–7.7 mm
- Family: Chrysomelidae
- Life cycle: One to three generations produced annually
- Range: North America east of Rocky Mountains
- Food: Adults and larvae feed on many plants, especially cucumbers and corn

Spotted Cucumber Beetle
Diabrotica undecimpunctata howardi

Spotted cucumber beetles are commonly found on many kinds of garden flowers and vegetables, and they are also occasionally attracted to lights.

IDENTIFICATION The bodies of these beetles are oval, moderately shiny, and mostly yellow or greenish yellow above, marked with bold spots. The head and all but the bases of the **antennae** are black. The unmarked **pronotum** is sometimes a slightly different shade than the **elytra,** each of which have three pairs of black spots. Underneath, they are mostly yellow, except for the black **metathorax** and mostly black legs.

HABITS A significant agricultural pest of North America, both adults and larvae of this species attack cucumbers, corn, sweet potatoes, and many other garden and crop plants. Overwintering adults emerge from plant debris in spring to feed and mate. They devour all parts of the food plant growing above ground, including even pollen and corn silks.

REPRODUCTION Eggs are laid in the soil near seedlings of target food plants. The pale, wormlike, brown-headed larvae, known as southern corn rootworms, attack the roots of corn and other plants, sometimes stunting their growth and killing seedlings. They take about a month to develop and pupate in earthen pupal chambers.

SIMILAR SPECIES The western spotted cucumber beetle, *D. u. undecimpunctata*, is widespread in western North America, where it can become an occasional and localized pest of fruit trees. Adults are very similar in appearance and habit to the spotted cucumber beetle, but the legs and underside are entirely black. The checkered melon beetle, *Paranapiacaba tricincta*, a southwestern species similar in size, has a black head, orange pronotum, and whitish elytra with somewhat rectangular black markings. This species occurs in the Plains and Rocky Mountain states.

Larva, known as the southern rootworm

Mosquitoes & Flies

Order Diptera

Adults have one pair of wings and knoblike **halteres** that serve as balancing organs during flight. Their mouthparts lap up plant and animal fluids or pierce animal tissues. Development is by **holometaboly.**

- Size: L 20.0–38.0 mm
- Family: Pediciidae
- Life cycle: One generation produced annually
- Range: Northeastern North America
- Food: Adults do not feed; larvae eat small adult and larval insects

Giant Eastern Crane Fly

Pedicia albivitta

Sometimes called giant mosquitoes or mosquito hawks, crane flies neither bite nor eat mosquitoes, and they are not closely related to these blood-sucking insects.

IDENTIFICATION Eastern crane flies are brown with white markings with a slender **abdomen** and long, slender legs with yellowish **femora,** dark at the tips. The short stem of the Y-shaped wing band reaches the wing's margin. The tip of the abdomen is pointed (males) or blunt (females). Crane flies are often mistaken for mosquitoes, but they lack piercing-sucking mouthparts and are not clothed in **scales.**

Larva

HABITS The conspicuous adults are commonly encountered in moist woodlands along shaded streams, rivers, and their tributaries, or near woodland pools. The large predatory larvae (45–60 mm long) are aquatic: They live in cold water and breathe air through a pair of **spiracles,** or breathing pore, on a disc located at the tip of the abdomen. They pupate beneath saturated mats of moss along the shore. Adults emerge from late spring through summer and are sometimes attracted to lights.

SIMILAR SPECIES Giant eastern crane flies are the most commonly encountered of the 60 species of *Pedicia* in eastern North America. The largest crane fly in the west, *Holorusia hespera,* is reddish brown with plain wings and white markings on the **thorax.**

- Size: L 5.0–7.0 mm
- Family: Culicidae
- Life cycle: **Multiple generations produced annually**
- Range: **Southeast Asia; established in eastern United States, South America, Europe, parts of Middle East, and Africa**
- Food: **Adults drink nectar; females take blood meals from birds and mammals**

Asian Tiger Mosquito
Aedes albopictus

Native to southeastern Asia, this boldly marked mosquito was accidentally introduced into North America in a shipment of used tires in Houston, Texas, in 1985.

IDENTIFICATION Tiny **scales** form the distinctive black-and-white stripes and bands on the body and legs of this species, and the top of the **thorax** has a single white stripe down the middle. Asian tiger mosquitoes have long, slender piercing-sucking mouthparts. Slightly smaller males have bushier **antennae** than females.

HABITS Both sexes drink nectar for their own nourishment, and females draw blood from their hosts, including humans, on spring and summer days to obtain nutrients necessary for egg development.

REPRODUCTION Eggs are laid individually near any body of stagnant water. Used tires kept outdoors are often selected, but these insects also lay eggs in tree holes, birdbaths, water-filled flower pots, blocked rain gutters, trash heaps, clogged storm drains, and ditches. Aquatic larvae, or wrigglers, hatch in a few days, breathe from underwater through siphonlike **spiracles** located near the tip of the **abdomen,** and eat bits of submerged vegetation. Active, comma-shaped pupae are called tumblers; they obtain oxygen through horn-shaped respiratory tubes on the thorax.

SIMILAR SPECIES The yellow fever mosquito, *A. aegypti,* is paler and has a lyre-shaped pattern of white stripes on top of the thorax. It occurs mostly in the southeastern United States.

Of Interest Female *Aedes* mosquitoes are **vectors** of several viral pathogens that cause yellow and dengue fevers, chikungunya, West Nile, and Zika viruses in humans, as well as parasitic heartworms that sicken dogs and cats.

Aquatic larva

TAKE THE BITE OUT OF MOSQUITOES

Mosquito bites not only make us itch; they can also make us sick. Chikungunya, dengue, West Nile fever, and Zika are just some of the mosquito-borne diseases present in North America, and malaria and yellow fever infections occur not far south of the United States. Whether you are at home, taking a day hike, on summer vacation, or traveling abroad, it is important to use repellent and dress appropriately to prevent discomfort and the spread of disease.

For information on avoiding mosquito bites and mosquito-borne diseases, visit the Centers for Disease Control website: www.cdc.gov.

WEAR INSECT REPELLENT

For protection against mosquitoes, use repellents that contain DEET, picaridin, IR3535, oil of lemon eucalyptus, or PMD (para-menthane-3, 8-diol). Products containing 30 to 50 percent DEET provide the longest protection, while those containing 15 percent picaridin and other active ingredients must be

Neglected pools serve as breeding sites for mosquitoes that may carry diseases like West Nile virus.

Asian tiger mosquitoes are a nuisance not only to humans but also to dogs, cats, and other animals.

applied more frequently. Always read the product directions carefully, paying close attention to warnings, especially regarding its use on infants, young children, clothing and gear, or with sunscreen. Apply as directed and avoid getting repellent in your eyes or mouth.

COVER UP

Sweat and bare skin attract hungry female mosquitoes. Keep cool and covered by wearing lightweight, loose-fitting, and light-colored long pants, long-sleeved shirts, and hats. Treat clothing with 0.5 percent permethrin, following product instructions carefully. Never apply permethrin directly on the skin. Treated articles remain repellent for several washings. Clothing pretreated with permethrin sold through various outlets will repel mosquitoes (and ticks) after many washings.

KEEP MOSQUITOES OUTSIDE

Secure your home by installing or repairing window screens, sealing doors, and making sure that pet doors work properly so mosquitoes cannot get inside. Eliminating points of ingress for mosquitoes and other insects also conserves energy and saves money on cooling and heating bills.

ELIMINATE BREEDING SITES

Bug zappers are ineffective against mosquitoes and kill mostly beneficial and harmless insects. Blanket application of insecticides needlessly kills all insects and harms other wildlife, too. One of the most effective ways to reduce mosquito populations is to destroy their breeding sites. All it takes is a bit of soil, compost, or decaying leaves mixed in about an inch of standing water, and in 7 to 14 days, voilà, you have mosquitoes! Neglected birdbaths, pools, and fishless ponds produce mosquitoes. Small containers, buckets, dishes beneath flower pots, gardening equipment, toys, stopped-up rain gutters, and covered drains outside basements also make excellent mosquito breeding grounds. Mosquito dunks contain bacteria that kill mosquito larvae only and are used in breeding sites that can't be drained, such as nearby catchment basins and storm drains.

- Size: L 15.0–19.0 mm
- Family: Bombyliidae
- Life cycle: One generation produced annually
- Range: Southern Ontario and eastern United States
- Food: Adult food is unknown; larvae are parasitoids of eastern carpenter bees

Tiger Bee Fly
Xenox tigrinus

This large fly with boldly marked wings is often seen in summer hovering singly or in small groups around fences, railings, signs, and other wooden structures.

IDENTIFICATION They are dark brown or black with large **compound eyes,** their bodies clothed mostly in short, dark hair-like **setae,** with scattered patches of white setae, especially near the sides of the **abdomen.** The large wings are held out to the sides of the body and have dark spots and irregular patches that coalesce to form a distinctive pattern.

HABITS Tiger bee flies are found in disturbed areas and deciduous forests—wherever eastern carpenter bees make their nests, especially near structures with exposed timbers, open habitats, and woodland edges.

REPRODUCTION Adults emerge in summer and mate end-to-end. Females lay their eggs at the entrance of carpenter bee nests occupied by pupae or adults. They hatch, overwinter, and infest the bees' newly provisioned nests in the spring. The fly larva parasitizes and eventually kills the developing bee, and then pupates within the nest.

SIMILAR SPECIES All five species of *Xenox* in Canada and the United States are parasitoids of carpenter bees (*Xylocopa*), and all but *X. tigrinus* occur in the southwestern United States and northern Mexico. Other similar large bee flies include species in the genus *Anthrax*; they tend to have hairier bodies and wings with more speckles than the tiger bee fly.

Of Interest The pupa has bristles that help anchor it inside the bees' nest tunnels so that the emerging fly can wriggle free. The empty pupal "skins" are left behind, protruding from the nest's entrance.

Carpenter bee at nest entrance, location for laying eggs

- Size: L 3.0–6.0 mm
- Family: Dolichopodidae
- Life cycle: Unknown
- Range: Southeastern Canada and United States
- Food: Adults and larvae prey on small invertebrates

Long-Legged Flies
Condylostylus species

These slender, leggy flies run across leaves and flit about lush, sun-dappled vegetation in gardens, parks, forest edges, and other moist habitats in late spring and early summer.

IDENTIFICATION Bright metallic green or blue, sometimes with brassy reflections, they have large reddish or greenish **compound eyes** broadly separated by a distinct notch. Within this notch is a pair of **tubercles** bearing erect, hairlike **setae.** The two clear wings sometimes have a dark U-shaped mark that touches the leading edge of the wing. The slender legs are dark or pale. The long, tapered **abdomen** is curled under at the tip in males.

HABITS Adults prey on small, soft-bodied arthropods, including aphids, whiteflies, blackflies, mites, and other flies. Abundant in agricultural systems,

Condylostylus and other long-legged flies may be important **predators** of insect pests. The vibrant, metallic colors of this insect may serve to warn hungry predators of their ability to quickly evade capture, thus saving energy for both prey and predator.

SIMILAR SPECIES The genus *Condylostylus* is found worldwide in the tropics and is most diverse in the neotropical region. Thirty-three species are found north of Mexico, all but three east of the Rockies. They are distinguished from other shiny, slender, metallic-green, long-legged flies by the broad notch between the compound eyes and the erect setae in it.

Of Interest Entomologists and photographers attempting to capture images of these flies can attest to their lightning-fast reflexes. Studies show that they can react to a camera flash in as little time as two to five milliseconds.

Adult with insect prey

- Size: L 15.0–20.0 mm
- Family: **Stratiomyidae**
- Life cycle: **Several generations produced annually**
- Range: **Widespread in warm climates worldwide**
- Food: **Adults do not feed; larvae eat decaying plant and animal tissues**

Black Soldier Fly

Hermetia illucens

These sluggish, wasplike flies gather at compost heaps and other accumulations of decaying plants in gardens, parks, and farms, where they mate and lay eggs.

IDENTIFICATION This species has relatively long **antennae,** dark smoky wings folded flat over the back at rest, and pale **tibiae** and **tarsi.** The translucent patches at the base of the **abdomen** inspired the scientific name—*illucens,* "enlightening"—and suggest this species' other common name: window fly.

HABITS Adults cannot bite or sting and are not attracted to human foods like some other flies.

REPRODUCTION Females lay masses of up to 500 eggs near carrion, dung, garbage, and other decaying organic matter, where the larvae feed and develop. The somewhat flattened larvae have

tough, leathery, and bristly **exoskeletons.** Available commercially as live food for captive amphibians, reptiles, and tropical fish, these larvae have been dubbed "Phoenix worms," "reptiworms," or "calcigrubs." Black soldier fly larvae are also used commercially to help break down compost and sanitize waste. Poultry and swine farmers sometimes employ their services in manure management programs to consume animal waste and destroy the eggs of pestiferous house flies and blow flies.

SIMILAR SPECIES These flies may resemble pipe organ mud dauber wasps (p. 185), but they can be easily distinguished because they have only two wings and all of their tibiae and tarsi are pale in color.

Of Interest Sometimes found in association with human remains, black soldier fly larvae and other insects are useful to forensic entomologists for establishing the time of death.

Larva

- Size: **L 12.0–15.0 mm**
- Family: **Syrphidae**
- Life cycle: **Two or three generations produced annually**
- Range: **Cosmopolitan**
- Food: **Adults sip flower nectar; larvae eat decaying organic matter**

Drone Fly
Eristalis tenax

*Drone flies, originally from Europe and now distributed nearly worldwide, are **Batesian mimics** of the European honey bee but are incapable of stinging or biting.*

IDENTIFICATION They are beelike in shape and color and have a broad head, large **compound eyes,** two wings, a broad waist, and a dark **abdomen** with golden brown triangles at its base. In males the compound eyes are narrowly separated on the top of the head; in females they are farther apart.

HABITS Drone flies are found in various open and disturbed habitats, including croplands. They are fond of visiting flowers in the aster family, especially in late summer, and are considered important pollinators.

REPRODUCTION Females lay up to 200 eggs near stagnant water or in decaying organic tissues and liquid excrement in stockyards, outdoor toilets, and septic tanks. Their sausage-shaped larvae, known as rat-tailed **maggots,** have a long, slender respiratory siphon located on the tip of the abdomen. The tip of this breathing structure just reaches the water surface, allowing the larvae to remain submerged as they feed on decaying animal waste and other organic matter in fetid pools with little dissolved oxygen.

SIMILAR SPECIES Most of the 27 species of *Eristalis* that inhabit Canada and the United States are native. Drone flies are distinguished from these other beelike flies by the two narrow bands of hairlike setae across each compound eye.

Of Interest Drone fly larvae can also develop in juicy carcasses, and they likely inspired ancient rituals based on the myth that honey bees spontaneously appeared from dead oxen, cows, and other animals.

Larva with respiratory tube extended

- Size: L 5.0–8.0 mm
- Family: **Muscidae**
- Life cycle: **Multiple generations produced annually**
- Range: **Cosmopolitan**
- Food: **Adults imbibe liquids associated with various plant- and animal-based foods**

House Fly
Musca domestica

An Afrotropical and Oriental native, the nonbiting house fly probably came to North America with European colonists.

IDENTIFICATION They have reddish **compound eyes,** four black stripes on a gray **thorax,** two clear wings held at an angle over the body, and a yellowish and gray **abdomen.** The compound eyes are narrowly (males) or widely (females) separated.

HABITS House flies are always associated with humans. They inhabit neighborhoods, parks, and garbage dumps, and they are especially abundant on farms and ranches and in stables. They land on foodstuffs, liquefy them with salivary secretions, and sop them up with their fleshy, spongelike mouthparts. Because of their feeding habits and contact with filth, the adults are potential carriers of pathogens that cause diarrhea, typhoid, cholera, and eye infections or that facilitate the transmission of parasitic worms.

REPRODUCTION Eggs are laid singly or in batches. The **maggots** develop in manure, lawn clippings, compost heaps, and other accumulations of wet organic matter. They breathe through a pair of D-shaped **spiracles** located on the tip of the abdomen. The entire life cycle, from egg to adult, may take as little as eight days. Adults typically live about two weeks. They breed year-round across the southern United States.

SIMILAR SPECIES Face flies *(M. autumnalis)* are similar: The females sponge up secretions on the faces of cattle and horses. Stable flies *(Stomoxys calcitrans)* have a distinct **proboscis** and can inflict painful bites.

Of Interest Cooler temperatures and access to food sources that are high in sugar are associated with longer life spans for house flies.

Larva (maggot)

- Size: L 10.0–14.0 mm
- Family: **Calliphoridae**
- Life cycle: **Multiple generations produced annually**
- Range: **Cosmopolitan**
- Food: **Adults drink plant and animal fluids; larvae eat decaying flesh**

Green Bottle Fly

Lucilia sericata

Shiny green bottle flies are commonly seen around dog feces, garbage cans, and carrion on warm days.

IDENTIFICATION Usually bright metallic green or occasionally coppery, these flies have brick-red eyes, an unmarked **thorax**, mostly clear wings with yellowish bases and brown veins, and a dull or shiny **abdomen.** The **antennae** and legs are black.

HABITS Green bottle flies feed on animal fluids and are among the first insects to arrive at a carcass.

REPRODUCTION Females lay masses of up to 200 eggs at a time on carrion, garbage, excrement, and other necrotic or decaying tissues, including infected open wounds. The developing **maggots** feed on decomposing tissue and complete their development in 3–10 days, depending on temperature and food quality. The developmental stages

of the maggots present in a corpse, used in combination with local weather conditions, help investigators determine the time of death.

SIMILAR SPECIES *Lucilia cuprina* is usually coppery and can be confused with similarly colored *L. sericata.* The black blow fly *(Phormia regina)* is dark metallic blue or green with bright-orange **setae** surrounding the first pair of **thoracic spiracles.** The blue bottle fly *(Calliphora vomitoria)* is larger and blue with yellow or orange setae on the head.

Of Interest When surgery and antibiotics are ineffective, doctors sometimes treat diabetic ulcers and other deep-tissue wounds with green bottle fly maggots raised under sterile conditions. The maggots eat only dead, infected tissue and produce antibacterial secretions that hasten the healing process, especially in patients infected with MRSA.

Larva (maggot)

- Size: L 8.0–14.0 mm
- Family: Sarcophagidae
- Life cycle: Several generations produced annually
- Range: Cosmopolitan
- Food: Adults drink animal fluids; larvae eat decaying flesh

Flesh Fly
Sarcophaga africa

Adult flesh flies prefer open and sunny habitats and are quick to find **feces** *and carcasses upon which to feed and reproduce.*

IDENTIFICATION They resemble an oversize and strikingly marked house fly, with large brick-red eyes, three black stripes on the **thorax,** clear wings, and a distinct black-and-white checkerboard pattern across the **abdomen,** the tip of which is reddish in males.

HABITS This fly is known to cause accidental intestinal infections in domestic animals. Females may deposit larvae on feed that is later consumed by an animal, causing it temporary intestinal distress. Attracted to fecal odors, females occasionally fly into barns or stables to deposit their larvae on the rectal membranes of immobile animals. **Maggots** may also enter the body through the nose and ears.

REPRODUCTION Females brood their eggs internally and then deposit first **instar** maggots ready to feed on rapidly decomposing feces and flesh, including human remains. Forensic entomologists use their presence on a corpse to calculate the **postmortem interval** and determine the time of death. The pale yellowish maggots complete their development in less than a week and pupate in the soil. Adults emerge in about two weeks and in a few days begin mating multiple times with the same fly or with different partners.

SIMILAR SPECIES Flesh fly species in several genera superficially resemble one another and can be reliably identified only by microscopic examination of the male's genitalia and other physical characteristics.

Of Interest Forensic entomology is the science of using insects in criminal investigations. The species of larvae found can help pinpoint the time of death up to a month later.

Larva

- Size: L 12.0–13.0 mm
- Family: Tachinidae
- Life cycle: Unknown
- Range: British Columbia and Montana to California
- Food: Adults sip flower nectar

Tachinid Fly
Adejeania vexatrix

One of the most striking tachinid flies in North America, few details of its biology are known other than the adults' late summer visits to milkweeds, thistles, and other mountain flowers.

IDENTIFICATION This large, robust, and bristly fly has long and slender mouthparts, brick-red **compound eyes,** mostly yellowish-brown head and **thorax,** and a pair of slightly brownish translucent wings. The broad, shiny **abdomen** is medium to dark orange with one or more black spots down the middle and covered mostly in golden bristles. Long black bristles are scattered across the top of the thorax but are concentrated along the lines between abdominal segments.

HABITS All known tachinid larvae complete their development within an insect host, ultimately killing it in the process, so it is likely that the larvae of this species are **parasitoids** of **caterpillars** or some other insect. Most tachinids attack caterpillars, beetles, true bugs, and sawflies, while others prefer grasshoppers, earwigs, and other arthropods. Tachinid flies can be generalists or specialists in terms of their host preferences and are sometimes found useful in controlling crop pests. They do not bite or spread disease, and they are not attracted to food or waste.

SIMILAR SPECIES In the West and Southwest, *Hystricia abrupta* has abdominal bristles that are scattered across the surface and lacks conspicuously long mouthparts. *Paradejeania rutilioides* has a mostly gray head and thorax, inconspicuous mouthparts, and a large, mostly yellowish-brown abdomen densely covered with black bristles.

Of Interest Adult tachinid flies feed on pollen and nectar and can serve as important pollinators, especially in places where bees are scarce.

Milkweed woolly bear, larval host of *Hystricia abrupta*

Butterflies, Skippers & Moths

Order Lepidoptera

Lepidopterans have four wings covered in **scales.** Adults usually have a coiled **proboscis** for sipping fluids, while their mostly herbivorous **caterpillars** have chewing mouthparts. Development is by **holometaboly.**

- Size: W 70.0–86.0 mm
- Family: Papilionidae
- Life cycle: **Up to three generations produced annually**
- Range: **Southern Ontario; eastern United States to Arizona; Oregon and northern California**
- Food: **Adults sip nectar; larvae eat pipevine leaves**

Pipevine Swallowtail

Battus philenor

Pipevine swallowtails flutter from flower to flower and sip nectar through a long proboscis.

IDENTIFICATION The upper wing surfaces of these tailed butterflies are mostly dark, with the male's hind wings more iridescent bluish green than those of the female. Hind wings of both sexes have distinct **submarginal** spots above and mostly iridescent blue-green undersides with orange spots and white spots along the margins. Mature **caterpillars** are purplish or red with two long, fleshy tentacles behind the head, six more tentacles at the end of the **abdomen,** and pairs of fleshy orange **tubercles** down the back.

HABITS Pipevine swallowtails fly spring to fall, sipping nectar from pinkish and purplish flowers. The caterpillars sequester toxins from pipevine plants into their tissues, protection from **predators** that carries through to the adult stage.

REPRODUCTION Male scent glands induce females to mate. Reddish-brown eggs are deposited under leaves and on stems of pipevines especially along streams and rivers.

SIMILAR SPECIES Several palatable butterflies mimic distasteful pipevine swallowtails, including the dark-form female eastern tiger swallowtail (*Papilio glaucus*) and both sexes of the spicebush (*P. troilus*) and black (*P. polyxenes*) swallowtails. The lookalike red-spotted purple (*Limenitis arthemis astynax*) lacks hind wing tails.

Caterpillar

- Size: W 70.0–100.0 mm
- Family: Papilionidae
- Life cycle: Up to three generations produced annually
- Range: Mostly west of Rocky Mountains
- Food: Adults drink nectar; larvae eat leaves of deciduous trees

Western Tiger Swallowtail
Papilio rutulus

Familiar residents of urban and suburban gardens and parks, these butterflies more typically inhabit winding watercourses and wooded canyons from the Rocky Mountains to the West Coast.

IDENTIFICATION These swallowtail butterflies have bold black stripes on their forewings and a single tail on each hind wing. The underside of each forewing is marked with a continuous yellow **submarginal** band. Mature **caterpillars** are mostly green with two conspicuous eyespots on their swollen **thorax,** followed by a yellow band.

HABITS Western tiger swallowtails fly mostly in early spring and summer, occasionally to late fall. Seeking nectar, they visit many kinds of flowers. They sometimes congregate in large numbers at mud deposits alonside streams or mud puddles in roads and yards.

REPRODUCTION Females lay their green spherical eggs on many riparian hardwoods, especially alder, cottonwood, willow, and sycamore. Mature caterpillars rest in shelters of leaves drawn together with silk produced by glands in their mouthparts. The tip of the dark brown pupa's **abdomen** is attached to a button of silk affixed to a vertical stem or surface and supported in an upright position by a heavy strand of silk. Two or three generations are produced annually in coastal areas and valleys, but only one summer generation occurs at higher elevations. The season's final generation overwinters as pupae that emerge in spring to feed, mature, mate, and reproduce.

SIMILAR SPECIES In the mountain canyons of the West, the two-tailed swallowtail *(P. multicaudata)* has narrower black wing stripes and a larger and smaller tail on each hind wing, while the wings of the widespread pale swallowtail *(P. eurymedon)* are much paler above and below.

Caterpillar

- **Size:** W 63.0–110.0 mm
- **Family:** Papilionidae
- **Life cycle:** Up to three broods produced annually
- **Range:** Eastern North America
- **Food:** Adults sip nectar; larvae eat leaves of plants in parsley family

Black Swallowtail
Papilio polyxenes

Black swallowtails fly from late spring through summer, visiting clover, milkweed, thistle, ironweed, joe-pye weed, and other flowers for nectar.

IDENTIFICATION The males are mostly black above with yellow spots along the edges and a broad yellow **submarginal** band. In females the yellow bands are reduced to small spots on the forewings and blue **scales** on the hind wings. The hind wings of both sexes have a red-and-black eyespot. The young black-and-white larvae resemble bird droppings, while mature **caterpillars** are mostly green with black bands and yellow spots.

HABITS Found in parks and gardens, seeking summer flowers, they will also gather at mud. Males patrol for females from hilltops and other prominent perching sites. Sometimes called parsley worms, the caterpillars are occasionally considered nuisances to gardeners growing carrots, dill, fennel, and parsley.

REPRODUCTION Courtship involves briefly fluttering about one another before landing to mate. Yellowish or cream eggs are laid singly on the leaf tips of caterpillar food plants related to parsley, including natives, weedy exotics, and cultivated plants. Two broods are produced annually, with sometimes a third in warmer regions.

SIMILAR SPECIES This is one of several eastern butterfly species that mimic the toxic pipevine swallowtail. A mostly yellow subspecies (*P. polyxenes coloro*) and the similar yet uncommon indra swallowtail (*P. indra*) are the only swallowtails found in the arid mountains of the Mojave, Colorado, and Sonoran Deserts.

Of Interest When threatened, black swallowtail caterpillars reveal a forked, orange **thoracic** gland that releases a noxious odor smelling like spicy vomit.

Caterpillar

- Size: W 32.0–48.0 mm
- Family: Pieridae
- Life cycle: Up to four or more generations produced annually
- Range: Europe and North America
- Food: Adults drink nectar; larvae eat plants in mustard family

Cabbage White
Pieris rapae

A European species first reported from Québec in 1860, the cabbage white is now widespread throughout most of North America.

IDENTIFICATION This mostly white butterfly has black-tipped forewings, each with one (male) or two (female) black spots. The undersides of the hind wings are pale yellow, white, or grayish white and unmarked in both sexes. Bright-green **caterpillars** have thin yellow stripes down the back and sides and are clothed in short, fine **setae.**

HABITS One of the first spring butterflies, they fly low in search of flowers, nectar, and egg-laying sites in gardens, parks, vacant lots, agricultural fields, and other open habitats supporting crucifers. Dandelions and various mustards are favored nectar sources in the spring, while clovers, mints, and asters are sought later in the season.

REPRODUCTION Eggs are laid singly on the young growth of mustards, radishes, cabbage, broccoli, cauliflower, and turnips. The green or tan chrysalis is strongly tapered on both ends, with projections on both sides. The tip of the **abdomen** attaches to a button of silk affixed to a leaf or other surface, the chrysalis suspended upright by a strand of silk.

SIMILAR SPECIES The mustard white (*P. napi*), a typically unmarked and montane species that prefers forest and tundra, sometimes has forewings with dark tips and one or two spots. Greenish veins on the undersides of their hind wings in spring become faint in later broods. The unmarked West Virginia white (*P. virginiensis*) flies in spring, is uncommon, and prefers moist woods.

Of Interest Cabbage whites are strong fliers, and researchers estimate that they may fly more than 100 miles over the course of their short lifetimes.

Caterpillar

- Size: W 23.0–32.0 mm
- Family: Lycaenidae
- Life cycle: Up to four generations produced annually
- Range: Across southern Canada and United States
- Food: Adults sip nectar; larvae eat plants in pea and mallow families

Gray Hairstreak
Strymon melinus

From spring to fall, gray hairstreaks are commonly seen visiting all kinds of flowers in gardens, parks, old fields, edges, and other open habitats.

IDENTIFICATION The upper wing surfaces are slate gray with a distinct reddish-orange eyespot near the long tail on each hind wing, while the mostly plain blue gray undersides have narrow, broken, white and black bands sometimes edged in orange. The hairlike tails and eyespots on the hind wings suggest false **antennae** and eyes that presumably direct **predator** attacks away from more vital body parts. Males are easily distinguished from females by their bright-orange **abdomens.** Mature **caterpillars** are bright green, pale yellow, reddish brown, or pink. Covered with stiff **setae,** they are **cryptically** and variably marked with white, cream, or light purple stripes on the sides.

HABITS Unlike other hairstreaks, gray hairstreaks sometimes rest with wings open. Males claim perches low on herbaceous growth and shrubs, especially on hilltops, to await passing females and will vigorously chase rival males away.

REPRODUCTION Pale green eggs are laid singly on young leaves and flower buds, especially clover, milkvetch, lupine, and mallow. Young caterpillars eat flowers and fruit, while older larvae may consume leaves. The last generation produced for the season overwinters as a pupa. One or two generations are produced in the north, three or four in the south.

SIMILAR SPECIES The southeastern red-banded hairstreak (*Calycopis cercrops*) has hind wings partly blue above and a thick red band underneath.

Of Interest Gray hairstreak caterpillars excrete **honeydew** and are tended and defended by ants.

Caterpillar

- Size: W 64.0–95.0 mm
- Family: Nymphalidae
- Life cycle: Multiple generations produced annually
- Range: Southern third of United States
- Food: Adults sip nectar; larvae eat passionflower leaves

Gulf Fritillary
Agraulis vanillae

Gulf fritillaries are sometimes abundant in gardens and parks, especially where passionflowers grow.

IDENTIFICATION Their wings are bright orange with bold dark markings above and silvery spots below. The upper surfaces of their long forewings have dark markings and three white spots ringed in black. Hind wings have broad dark margins with orange spots above and brown with long silvery spots below. Females are larger, somewhat darker, and have more extensive markings. Mature **caterpillars** can be blackish, dark brown, or purplish with orange and white stripes, or they can be completely orange. All have six rows of branched black spines running down their back and sides.

HABITS Gulf fritillaries fly quickly and erratically in search of mates and nectar. They are particularly fond of red or white flowers. One or more may roost together on

vines and vertical branches at night. Gulf fritillaries migrate northward during the summer, especially in California, in the Mississippi River Valley, and along the Atlantic coast, and they establish temporary breeding colonies all along the way. Intolerant of cold conditions, they will migrate southward to southern Florida to spend the winter, sometimes in large numbers. They can be found flying year-round in the warmer surrounds of southern California, southern Texas, and Florida.

REPRODUCTION This species' bright yellow eggs are laid singly on various parts of passionflower plants, which represent their caterpillars' sole food source. The irregularly shaped grayish chrysalis is suspended upside down from a button of silk affixed on the underside of horizontal branches.

SIMILAR SPECIES Gulf fritillaries resemble fritillaries (*Speyeria*) but are distinguished by their long forewings, with three black-encircled white spots above and large elongate silvery spots underneath.

Caterpillar

- Size: W 57.0–101.0 mm
- Family: Nymphalidae
- Life cycle: One or two broods produced annually
- Range: North America, less common in Southeast; Eurasia
- Food: Adults drink nectar, sap; larvae eat leaves of hardwoods

Mourning Cloak
Nymphalis antiopa

Among the first butterflies to appear in spring, mourning cloaks soar over gardens and parks and along roads, riparian woodlands, and forest edges.

IDENTIFICATION Their wings above are dark brown with yellow margins and blue spots. Underneath they have dark brown striations and pale yellowish margins. The mature, spiny black larvae have red or orange-red spots down their backs.

HABITS The long-lived adults emerge in summer and spend most of their lives avoiding the heat, becoming active again briefly in the fall. They appear to engage in regular seasonal movements, up and down mountain slopes and along beaches. As winter approaches, they seek shelter under loose bark or in woodpiles and sheds to avoid freezing. Mourning cloaks overwinter as adults and may emerge briefly on warmer days in late winter, even while snow is still on the ground, to bask in the sun, using their dark wings as solar collectors to gather heat and transfer it to their flight muscles.

REPRODUCTION Females lay clusters of pale green eggs on twigs that soon fade to white and then darken just before hatching. Young **caterpillars** are gregarious and feed within silken webs on the leaves of elm, hackberry, willow, and many other deciduous trees and shrubs. The spiny grayish chrysalis hangs upside down from a button of silk affixed on the underside of horizontal branches.

SIMILAR SPECIES Large dark butterflies with yellow wing margins, mourning cloaks cannot be confused with any other butterfly in North America.

Of Interest Mourning cloaks have one of the longest known life spans of any butterfly, living approximately 11 to 12 months.

Caterpillar

- Size: W 44.0–76.0 mm
- Family: **Nymphalidae**
- Life cycle: **Two generations produced annually**
- Range: **Widespread in Northern Hemisphere**
- Food: **Adults feed mostly on fluids other than nectar; larvae eat nettle leaves**

Red Admiral
Vanessa atalanta

Male red admirals stake out hilltops, shrubs, sheds, bare patches of ground, and other landmarks in anticipation of females passing by.

IDENTIFICATION Bold orange-red bands on the upper surfaces of the dark brown wings cross the forewings and skirt the outer margins of the hind wings. The tips of the upper surfaces of the forewings are marked with white spots, while the hind wing bands have black and metallic spots. Underneath, the forewings have white, black, and blue markings with a pinkish bar, while the hind wings are **cryptically** marbled with gray, brown, and black markings. Mature **caterpillars** range in color from almost white to yellow-green or black with pale flecking, and they have a dark head and pale, branched spines. Some individuals have creamy blotches on the sides that coalesce into a broad stripe.

HABITS Adults emerge in early and late summer and fly rapidly and erratically through gardens, parks, and woodlands, and along forest edges. They are mostly attracted to dung, carrion, sap, and fermenting fruit, but they are also known to occasionally visit wildflowers including thistles, milkweed, dogbane, clover, and Queen Anne's lace.

REPRODUCTION Pale green eggs are laid singly on the upper leaf surfaces of larval food plants. The young, gregarious larvae eat nettles and build silk nests that envelop the nettles' shoot tips. Older caterpillars feed alone. They will chew through a leaf's **petiole,** reattach the drooping leaf with silk, and begin eating from inside the leaf tip.

Of Interest Red admiral caterpillars are often found on plants in the nettle family (Urticaceae), most of which have stinging hairs that may protect the plants from herbivorous animals.

Caterpillar

- Size: W 44.0–67.0 mm
- Family: Nymphalidae
- Life cycle: Multiple generations produced annually
- Range: North America
- Food: Adults sip nectar, mud; larvae eat leaves of plants in sunflower family

American Lady
Vanessa virginiensis

Territorial males stake out bare patches, hilltops, roadsides, and mud puddles, while females dart erratically over the ground as they search for egg-laying sites.

IDENTIFICATION The upper surfaces of the American lady's wings are orangish in color overall with black-and-white markings. The tips of their forewings are somewhat extended and rounded at the very end, and their hind wings have two large bluish eyespots underneath. Mature **caterpillars** are spiny and they vary in color, but they typically have thin, creamy bands on the front of each body segment as well as red spots at the bases of the branched spines along the sides.

HABITS American ladies are residents across the southern United States, but they are also known to migrate and to temporarily occupy northern states and southern Canada. They visit asters, dogbane, goldenrod, marigold, milkweed, and other low-growing flowers in parks, fields, utility rights-of-way, forest edges, and other foliage-rich open areas.

REPRODUCTION The green, ribbed, and barrel-shaped eggs of the American lady are laid singly on the upper side of leaves. Young caterpillars are solitary feeders and eat a wide variety of plants in the sunflower family, especially everlasting, pearly everlasting, and plantain-leaved pussytoes. They will occasionally utilize species in other plant families. All caterpillar stages tie up bunches of leaves with silk and feed from within the shelter.

SIMILAR SPECIES There are two similar and widespread *Vanessa* species in North America, both of which have multiple eyespots underneath their hind wings. The West Coast lady (*V. annabella*) has distinctly extended forewings with tips that appear clipped, while the painted lady (*V. cardui*) has rounded forewing tips that extend only slightly.

Caterpillar

- Size: W 42.0–70.0 mm
- Family: Nymphalidae
- Life cycle: Up to four generations produced annually
- Range: Eastern and southern United States, Pacific coast to Oregon
- Food: Adults sip nectar; larvae eat leaves of many plants

Common Buckeye
Junonia coenia

These butterflies visit gardens, parks, and other open areas, sometimes gliding short distances between wing beats, and bask on the ground with wings open, especially on cooler days.

IDENTIFICATION Each of their brownish forewings has a pair of short orange bars, a broad whitish bar, and a large eyespot with whitish margins on the inside. Each of their hind wings has both a large and a small eyespot. Undersides of the wings are grayish brown or tan in spring and summer, reddish brown in fall. Mature **caterpillars** have a head that is orange on top, branched metallic blueblack spines, and orange legs and **prolegs.** Variable in color, the caterpillars are usually dark above with pale or orange sides.

HABITS A resident of the southern United States, the common buckeye regularly migrates northward in summer, reaching as far as Oregon, New England, and southeastern Canada. Many flowers are sought as nectar sources, especially those in the sunflower family, such as thistles, daisies, and wild sunflowers. Males perch low to the ground as they wait for females to fly past. Adults live no longer than about 10 days.

REPRODUCTION The dark green eggs are laid singly on the upper surfaces of leaves. The solitary larvae eat the leaves of many different kinds of plants, especially those related to snapdragons and plantains. The chrysalis is variable in shape and location but always has rows of **tubercles** on the back and is cream colored with reddish blotches to black.

SIMILAR SPECIES In Florida, the mangrove buckeye (*J. everete*) has a large eyespot on the forewing surrounded by orange, while the tropical buckeye (*J. genoveva*) from southern Texas has a narrow band on the underside of the hind wing.

Caterpillar

- Size: W 86.0–124.0 mm
- Family: Nymphalidae
- Life cycle: Up to five generations produced annually
- Range: Across southern Canada and throughout United States
- Food: Adults sip nectar; larvae eat milkweed leaves

Monarch
Danaus plexippus

Large and beautiful, the monarch is the most familiar butterfly in North America and the official state insect or butterfly in seven states.

IDENTIFICATION Wings are burnt orange with black veins and white-spotted margins above, paler underneath. Male's hind wings have scent pouches, appearing as a symmetrical pair of dark smudges in the middle of each wing. The **caterpillars** have thick black, yellow, and white bands and a pair of tentacles on each end.

HABITS Monarchs are typically seen in gardens, parks, fields, and other developed or natural open habitats. They are of great scientific interest because they are among the few insects that regularly engage in annual two-way migrations. The last generation of butterflies begins migrating in late summer and early fall. Those east of the Rockies fly to protected sites in the mountains of southern Mexico, while butterflies in the West seek shelter along the Pacific Coast of California and Baja California. Coastal southern California and Florida have populations that do not migrate. In spring, surviving migrants leave their overwintering grounds to repopulate North America, producing three to five generations throughout the summer.

REPRODUCTION Greenish-white or cream eggs are laid singly under milkweed leaves. The caterpillars sequester the plant's distasteful chemicals in their own tissues as they feed. The caterpillars' bold and **aposematic** colors warn potential **predators** of their distastefulness. This strategy and chemical defense carries on through to the adult.

SIMILAR SPECIES The less boldly marked queen (*D. gilippus*) is smaller, while the hind wings of the viceroy (*Limenitis archippus*, opposite) have a narrow black band across the middle—the best way to distinguish monarchs from viceroys.

Caterpillar

- Size: W 63.0–86.0 mm
- Family: Nymphalidae
- Life cycle: Up to three generations produced annually
- Range: North America east of Pacific provinces and states
- Food: Adults sip nectar and mud; larvae eat mostly willow leaves

Viceroy
Limenitis archippus

With monarchs and queens, viceroys form part of a **Müllerian mimicry complex** *of butterflies, all superficially resembling one another and similarly defended by distasteful chemicals*

IDENTIFICATION Their brilliant rusty reddish-orange wings are contrastingly marked with black veins and broad black-and-white–spotted margins. The middle of each hind wing is crossed by the distinctive narrow black band.

Viceroys resemble monarchs (*Danaus plexippus*) in the North, but in the southern parts of their range, they more closely resemble queens (*D. gilippus*). All three share bold colors and markings that are **aposematic** and warn **predators** of their bad taste. Florida viceroys tend to be more reddish, while populations in the Southwest have less pronounced wing veins and the bands across the hind wings are faint and may even be lacking. Mature **caterpillars** have two long **thoracic** horns behind the head and are mottled white with brown or olive, resembling a bird dropping.

HABITS Adults take wing in spring and fly throughout the summer. They are usually found in meadows, moist shrubby areas, and open riparian habitats in deciduous woodlands. In circumstances where flowers are not available, viceroys will visit moist dung, carrion, and decaying fruit. Territorial males patrol small areas in search of mates, and they will often return to the same low perch after each sortie.

REPRODUCTION Viceroys copulate on the wing, with the female carrying the male. Pale-green or pale-yellow eggs are laid singly on the upper side of leaf tips on the caterpillar's food plants, including willow, cottonwood, and poplar.

SIMILAR SPECIES Monarchs and queens lack the narrow black band that crosses the hind wing of the viceroy.

Caterpillar

MONARCHS IN TROUBLE

More than a million North American monarch butterflies are thought to migrate every year. Those east of the Rockies fly to 14 known sites in the oyamel fir forests high in the mountains of southern Mexico, while those west of the Rocky Mountains seek shelter among dozens of oak, cypress, sycamore, and non-native eucalyptus groves that hug the Pacific coast of California and Baja California. The forests and groves where monarchs overwinter serve as "thermal blankets" that protect them from frost, rain, wind, and dehydration.

Large-scale agriculture in the Midwest has impacted the monarch's summer breeding grounds by severely limiting food sources for both caterpillars and adult butterflies.

THE BIG SQUEEZE

A tropical species adapted to reproducing in a temperate climate, monarchs are already walking an ecological tightrope under the best of circumstances just to make it through the winter. Over several decades, real estate development in California and illegal logging in Mexico have stripped away sizable

The boldly striped caterpillar of the monarch butterfly may be seen on the buds and foliage of milkweed, its only food, but as these native plants are disappearing, so are the monarchs.

portions of the monarch's thermal blanket, exposing countless migrating butterflies to the winter elements. Concurrently in the Midwest, large tracts of land have been planted with corn and soybeans to produce biofuels, transforming the monarch's summer breeding grounds into virtual deserts fit for neither **caterpillars** nor butterflies. To complicate matters, these genetically modified crops are designed to be resistant to herbicides, thus encouraging farmers to apply more weed killers to their fields. Milkweed, the sole food plant for monarch caterpillars, is particularly susceptible to these chemicals, so very little still grows in and along the edges of croplands. The perfect storm created by habitat loss and increased herbicide use has resulted in the loss of more than 80 percent of North America's monarchs since the 1990s. It is hard to imagine that such a revered and ubiquitous insect is in this much trouble.

WHAT YOU CAN DO

Efforts are under way to protect monarchs and reclaim tracts of agricultural land in the Midwest by planting lots of milkweed. A true recovery, however, will occur only if gardeners, park groundskeepers, zoo and botanic garden horticulturists, county road department crews, and other keepers of urban and suburban lands mobilize to create herbicide- and pesticide-free habitats that are suitable for monarchs and other wildlife. Replacing sterile lawns and other exotic plantings with indigenous milkweeds for the caterpillars and a range of native plants that will flower throughout the growing season to sustain butterflies will also go a long way toward supporting them as well as other wildlife.

Consult your state's native plant society to find out which native plant species are best suited for your region. Visit *monarchwatch.org* and *xerces.org* for more information on monarch conservation.

- Size: W 43.0–67.0 mm
- Family: Hesperiidae
- Life cycle: One or two broods produced annually
- Range: Across southern Canada and most of United States
- Food: Adults sip nectar and mud; larvae eat leaves of plants in pea family

Silver-Spotted Skipper
Epargyreus clarus

The silver-spotted skipper is one of the largest, most distinctive, and widely distributed skippers in North America.

IDENTIFICATION Viewed from above, the chocolate brown forewings have dull irregular bars, while the fringed hind wings have a narrow checked pattern. Viewed from beneath, the hind wings have large and conspicuous silvery white patches. Mature **caterpillars** have oversize, brownish-red heads with a pair of orange spots. Their pale yellow-green bodies are marked with thin, dark bands, and they have red legs and yellow **prolegs.**

HABITS Silver-spotted skippers appear in spring and have a rapid and jerky flight that makes them difficult to follow. They prefer to seek nectar from purple, red, pink, or blue flowers growing in gardens, parks, and prairies and along woodland edges. Males seeking females perch low or high on twigs or leaves and will chase any large insect that flies nearby.

REPRODUCTION Females lay bright-green eggs with red tops singly on the upper sides of leaves of the larva's food plants, including black locust, honey locust, false indigo, Chinese wisteria, and other woody and herbaceous legumes. Newly hatched larvae build shelters by cutting a section of a leaf and rolling it with the upper surface inside. As they grow, the caterpillars construct larger and larger shelters that consist of several leaves. They forcibly eject their waste from these shelters, presumably to avoid attracting parasitic wasps. The last generation overwinters as a pupa.

SIMILAR SPECIES Many skippers are notoriously difficult to identify, but their large size and silvery white patches under the hind wings make it relatively easy to distinguish the silver-spotted skipper from others.

Caterpillar

- Size: W male 17.0–36.0 mm
- Family: Psychidae
- Life cycle: One generation produced annually
- Range: Eastern United States
- Food: Adults do not feed; larvae eat scales and leaves of trees and shrubs

Bagworm
Thyridopteryx ephemeraeformis

Bagworms—also known as evergreen bagworms, common bagworms, or common basketworms—are one of the most damaging pests of trees in the East.

IDENTIFICATION Adult males are blackish and have distinctly tapered bodies, feathery **antennae,** no functional mouthparts, and blackish translucent wings that lack **scales.** They develop in conspicuous silken spindle-shaped bags festooned with bits of twigs, cedar scales, and leaves. The soft-bodied, **caterpillar-**like females are yellowish white and virtually hairless; lack mouthparts, wings, legs, and antennae; and are seldom seen.

HABITS Males emerge in fall to mate, are sometimes attracted to lights, and die within a few days.

REPRODUCTION Using **pheromones** to locate a wingless female still in her larval enclosure, the male inserts his **abdomen** through an opening at the lower end of the bag to mate. She lays up to 1,000 smooth, cylindrical eggs inside the bag and soon dies. Larvae hatch in spring and disperse to search for suitable food plants. They feed throughout the summer on juniper, arborvitae, southern red cedar, and pine, as well as many hardwoods including maples, elms, and willows. While feeding, they construct a protective bag that is initially carried upright but eventually hangs downward with only the head, **thorax,** and well-developed legs of the larva protruding from the opening at the upper end. Pupation occurs in summer within the bag.

SIMILAR SPECIES There are 26 species of bagworms north of Mexico.

> *Of Interest* Heavy infestations of bagworm larvae can cause severe defoliation, resulting in aesthetic damage or serious injury of entire stands of trees, especially conifers.

Pupating bagworms

- Size: W 22.0–44.0 mm
- Family: **Lasiocampidae**
- Life cycle: **One generation produced annually**
- Range: **Eastern North America**
- Food: **Adults do not feed; larvae eat leaves of plants in rose family**

Eastern Tent Caterpillar
Malacosoma americana

Eastern tent caterpillars are often abundant in spring, while the moths are seen only occasionally at lights in the summertime.

IDENTIFICATION The chunky, fuzzy moths are light to dark chocolate brown with two parallel, nearly white lines across each forewing. The mature larva has a continuous yellow stripe down the back and is variegated with black, deep blue, orange, and white markings on the sides.

HABITS Young **caterpillars** are gregarious and construct silken tents in tree crotches and occupy them only at night and during inclement weather. The tent grows with the caterpillars and may reach 2 feet across. They venture out during the day to feed on leaves of apple, cherry, hawthorn, plum, and other trees and shrubs, sometimes completely defoliating them. Mature larvae are solitary and construct their chalky white cocoons under bark, boards, and eaves of homes and outbuildings.

REPRODUCTION The moths emerge in summer, and females lay dark masses containing 150–350 eggs around tree twigs. The first **instar** larvae develop that summer, but they don't hatch until leaves begin to appear the following spring.

SIMILAR SPECIES Six species of *Malacosoma* occur in North America. Eastern tent caterpillars are distinguished by a yellow stripe. Mature western tent caterpillars (*M. californicum*), found across Canada and the northern United States, have a narrow black and white **dorsal** stripe. The widespread forest tent caterpillar (*M. disstria*) has a series of keyhole patterns down the back and makes silken mats, not tents, on limbs and trunks.

> *Of Interest* The gregarious larvae construct silken tents in branch crotches for protection and sometimes defoliate trees and shrubs.

Caterpillars in crotch of tree

- Size: **L 35.0–42.0 mm**
- Family: **Erebidae**
- Life cycle: **One to four generations produced annually**
- Range: **North America**
- Food: **Adults probably don't feed; larvae eat leaves of many deciduous trees and shrubs**

Fall Webworm
Hyphantria cunea

The silken nests of fall webworms festooning trees and shrubs are familiar sights in neighborhoods and parks and along highways in late summer and fall.

IDENTIFICATION The fuzzy-bodied moths are pure white in the northern part of their range but have dark wing spots in the South. The bases of the front legs are bright orange or yellow. Mature **caterpillars** are hairy and variable in color.

HABITS The young, gregarious larvae skeletonize the leaves of hundreds of woody plant species, including fruit trees, elm, hickory, maple, cottonwood, and willow, as well as some herbaceous plants. They cover leaves with silk as they feed, eventually enveloping entire branches. Mature caterpillars are solitary. The pupae overwinter within white flimsy cocoons secreted in bark crevices and wall cracks, under benches, or mixed among leaf debris on the ground. Moths emerge in early spring in the South or during summer in the northern parts of their range.

REPRODUCTION Eggs are laid in masses on the undersides of leaves. The larvae feed and develop inside the tent until ready to pupate. The northern black-headed form is yellowish or greenish with pale, hairlike **setae,** while the southern red-headed form is darker with reddish-brown setae.

SIMILAR SPECIES Eastern tent caterpillars *(Malacosoma americana)* are active in spring and build smaller nests around the crotches of trees that are used as shelters only, while gypsy moth *(Lymantria dispar)* caterpillars don't make nests at all.

Of Interest Fall webworms, native to North America, were accidentally introduced into Europe in the 1940s and now inhabit parts of Asia as well.

Caterpillar

- Size: W 95.0–155.0 mm
- Family: **Saturniidae**
- Life cycle: **One generation produced annually**
- Range: **Southern New England to Florida, west to Missouri and Texas**
- Food: **Adults do not feed; larvae eat leaves, deciduous trees, and shrubs**

Regal Moth
Citheronia regalis

*The **caterpillars** of regal moths, large in size and fearsome in appearance, are called hickory-horned devils and are harmless to humans and pets.*

IDENTIFICATION These very large, stout moths have orangish forewings with broad gray stripes and yellowish spots, while the hind wings are more orange than gray and spotless. Mature caterpillars, usually seen when ready to pupate, are spectacular in size and form, reaching up to 120 mm in length. They are green with six long, reddish orange and black-tipped **thoracic** horns and broad, oblique white bands on the sides just below most of the abdominal **spiracles.**

HABITS Males are attracted to lights more so than females, especially those near deciduous woodlands and mixed forests. Adults emerge from their cocoons on late summer evenings and remain quiet throughout the following day.

REPRODUCTION Males start flying at dusk to search for "calling" females, which release their attractant **pheromones** around midnight; copulation continues through the next day. Eggs are laid singly or in small batches on upper and lower surfaces of leaves at dusk. Young larvae resemble bird droppings and feed nut and fruit trees. Mature caterpillars pupate in loose soil without a cocoon.

SIMILAR SPECIES The smaller pine devil moth (*C. sepulcralis*) is mostly brownish gray and more narrowly distributed along the Atlantic and Gulf Coasts. *C. splendens sinaloensis* of southeastern Arizona is similar to *C. regalis* in size but its base color is reddish.

Of Interest The caterpillars fiercely thrash their heads and thoraxes about when disturbed.

Caterpillar

- Size: W 80.0–174.0 mm
- Family: Saturniidae
- Life cycle: One or more generations produced annually
- Range: Eastern United States to Nebraska and Texas
- Food: Adults do not feed; larvae eat leaves of deciduous trees and shrubs, also pine needles

Imperial Moth
Eacles imperialis

The imperial moth is one of the most spectacular insects in the East. Nocturnal in its habits, it is usually encountered at lights in woodlands and forests.

IDENTIFICATION This large, stout moth has yellow wings variably marked with pinkish, orangish, or purplish-brown spots. Males are smaller, yellow, and more heavily marked with purplish patches on the outer forewings than females. Mature **caterpillars** reach up to 80 mm and are variable in color, ranging from green to various shades of red or brown, but are always clothed in long silky **setae.** They also have enlarged spiny horns on the **thorax** and usually a distinct white patch over each abdominal **spiracle.**

HABITS Imperial moths have largely disappeared from many urbanized areas in the northeastern United States. Adults emerge from their cocoons on summer mornings at dawn. Males are strong fliers and take wing at night to track "calling" females, which release sexually attractant **pheromones** long after midnight. The larvae feed on a wide range of trees and shrubs, especially pine, cedar, oak, maple, box elder, sweet gum, sycamore, basswood, birch, and sassafras.

REPRODUCTION Copulating pairs meet at night and remain together through the following day. Females begin laying bright-yellow eggs singly or in small clutches on upper and lower surfaces of leaves at dusk. When mature, caterpillars pupate in loose soil without a cocoon, like *Citheronia* (opposite).

SIMILAR SPECIES The subspecies *Eacles i. pini* is restricted to pine forests across the northern Great Lakes Basin. Oslar's eacles, *E. oslari*, occurring only in southeastern Arizona and Mexico, has a bold, brown line across each wing.

Caterpillar

- Size: W 50.0–80.0 mm
- Family: **Saturniidae**
- Life cycle: **One to four generations produced annually**
- Range: **Eastern North America to Manitoba, southwestern Utah, and Texas**
- Food: **Adults do not feed; caterpillars eat leaves of trees and shrubs**

Io Moth
Automeris io

When disturbed, io moths suddenly open their wings to reveal large mammal-like eyespots on their hind wings, possibly to intimidate would-be **predators.**

IDENTIFICATION This robust moth has yellow (males) or brownish, reddish, or purplish-brown (female) forewings. The relatively plain forewings conceal hind wings, each with a single black eyespot with a blue-ringed white center. Mature bright-green **caterpillars** have stinging spines.

HABITS Both males and females are attracted to lights in various habitats, including wooded suburbs, deciduous woodlands, and thorn forests. Adults emerge late mornings or early afternoons during the summer and remain still until evening. The number of generations per year varies from one in the North to three or four in southern Florida and Texas.

REPRODUCTION Females release **pheromones** in the evening to attract males. Pairs mate for most of the next day, separating at dusk. The female lays small batches of oblong eggs flat on leaf surfaces and stems. The eggs hatch within one to three days of each other. Larvae feed on the leaves of a wide variety of deciduous trees and shrubs, including redbud, hackberry, cherry, pear, willow, and sassafras, as well as several non-native plants. Mature caterpillars pupate in papery cocoons hidden on the ground among leaf litter or in crevices.

SIMILAR SPECIES Seven species of *Automeris* occur north of Mexico, but only *A. io* is widespread in the East. *Automeris louisiana* from coastal Louisiana is seldom seen, while the remaining five species occur in the Southwest.

Of Interest This moth is named for Io, a priestess of Hera, wife of Zeus, in Greek mythology.

Caterpillar

- Size: W 100.0–150.0 mm
- Family: Saturniidae
- Life cycle: One or two generations produced annually
- Range: Southern Canada, most of United States, except Arizona and Nevada
- Food: Adults do not feed; caterpillars eat leaves of deciduous trees and shrubs

Polyphemus Moth
Antheraea polyphemus

The polyphemus moth is the most widely distributed species of giant silk moth in North America.

IDENTIFICATION The light brown wings each have a black eyespot ringed in yellow, and those on the hind wings are further ringed in black and blue. The male's feathery **antennae** are more developed than those of the female. The head of the mature bright-green **caterpillar** is withdrawn within the **thorax.** The body segments are distinctly convex on the back, and most have an oblique yellow line passing through the **spiracle.**

HABITS The adults are sometimes common in late spring and summer in parks, forests, and woodlands, especially in large stands of oak. Both sexes are attracted to light. When disturbed, they flap their large wings and flash their eyespots to drive off **predators.** They emerge from their cocoons in late afternoon from late spring through summer. Cocoons often incorporate leaves of the food plant and are sometimes suspended from a branch by a slender, straplike **peduncle.**

REPRODUCTION Mating begins late in the evening, and mated pairs remain together throughout the next day until they part at dusk. Females lay banded, off-white eggs singly or in short rows of two or three on leaves.

SIMILAR SPECIES More than 35 species of *Antheraea* are known, most occurring in Asia. The only other North American species, *A. oculea*, occurs in the mountains of Arizona and western New Mexico and has wing eyespots ringed in orange with extensive blue and black scaling.

Of Interest This moth is named for the one-eyed giant cyclops encountered by Odysseus in Greek mythology.

Cocoon and caterpillar

- Size: W 75.0–105.0 mm
- Family: **Saturniidae**
- Life cycle: **One or more generations produced annually**
- Range: **Eastern North America**
- Food: **Adults do not feed; caterpillars eat leaves of deciduous trees**

Luna Moth
Actias luna

The spectacular luna moth was first reported from Maryland in 1700, but it was not described as a species until 1758 by Swedish biologist Carl Linnaeus.

IDENTIFICATION Large, robust, and pale green or bluish green, the luna moth has **antennae** that are more (male) or less (female) featherlike. Each hind wing has a long and distinctive tail. In flight, the fluttering and expendable tails distract bats from attacking more vital body parts. The mature lime-green **caterpillar** has magenta spots and a pale-yellow stripe running along its sides just beneath the **spiracles.**

HABITS Luna moths are one of the most common giant silk moths in eastern North America, especially in the southern United States, where they are often found at lights in suburbs, parks, forests, and woodlands.

REPRODUCTION Adults typically emerge on late spring and summer mornings and start mating the next day just after midnight. The whitish eggs are mottled with brown and are laid singly or in small batches on the food plant. Larvae are solitary and somewhat sedentary feeders on white birch in the North and on hickory, walnut, sumac, persimmon, and sweet gum in the South. When ready to pupate, the caterpillars spin an irregularly shaped, papery brown cocoon that usually incorporates a leaf of their food plant at the base. There is one generation produced annually in the North and up to three generations in the South.

SIMILAR SPECIES Large and pale green with long-tailed hind wings, luna moths cannot be confused with any other moth species in North America.

Caterpillar

Of Interest The long, fluttering tails of luna moths disrupt the echolocation signals of hungry bats.

- Size: W 110.0–150.0 mm
- Family: **Saturniidae**
- Life cycle: **One generation produced annually**
- Range: **Eastern North America to Montana and Colorado**
- Food: **Adults do not feed; larvae eat leaves of deciduous trees**

Cecropia Moth
Hylaphora cecropia

Popular in natural history literature because of its large size and coloration, the cecropia moth is one of the largest moths in North America.

IDENTIFICATION Their wings have white, reddish, and tan bands and large black-lined crescents, while the forewing has a black eyespot with a blue crescent inside. Males sport better-developed feathery **antennae** than females. Mature **caterpillars** have frosted green bodies covered with orange, yellow, and blue knobs.

HABITS Adults are typically found at night at lights in suburbs, parks, woodlands, and forests. They emerge from their cocoons midmorning during late spring and summer, rest for a day, and then mate just before dawn over the next several days.

Caterpillar and cocoon

REPRODUCTION Males mate with multiple partners, but females copulate only once. Short rows of two to six eggs are laid on upper and lower surfaces of leaves. Caterpillars eat maple and box elder leaves, as well as cherry, plum, apple, birch, alder, dogwood, and willow. Mature larvae leave their food plant to seek a sheltered site to construct a slender or bag-type cocoon that is pointed on both ends and attached lengthwise to a stem.

SIMILAR SPECIES Four more species inhabit North America. *Hylaphora columbia gloveri* occurs in eastern California, Nevada, and the Rockies; *H. euryalus* the Pacific Coast; and *H. kasloensis* from western Canada to Idaho and Montana.

Of Interest The cecropia moth is one of several native moths in decline in eastern North America because of a parasitic fly (*Campsilura concinnata*) introduced from Europe to control gypsy moths (*Lymantria dispar*), a major forest pest.

- Size: W 90.0–135.0 mm
- Family: **Sphingidae**
- Life cycle: **One or two generations produced annually**
- Range: **Parts of southern Canada and most of United States**
- Food: **Adults sip flower nectar; larvae eat leaves of tomato and tobacco**

Five-Spotted Hawk Moth
Manduca quinquemaculata

The five-spotted hawk moth is probably best known to gardeners and farmers by its **caterpillar,** *the tomato hornworm.*

IDENTIFICATION Moths have gray forewings with black-and white markings. Fringe along the margins of both pairs of wings are uniformly gray. The **abdomen** typically has five, occasionally six, pairs of yellow spots. Mature caterpillars have eight L-shaped lines on each side and a green horn with black edges.

HABITS Moths emerge from their pupae during the day and become active at dusk in summer. Both sexes are attracted to lights, sometimes in large numbers, in a wide variety of habitats, especially near tobacco and tomato crops. They visit flowers of periwinkle, Japanese honeysuckle, petunia, phlox, and desert tobacco.

Caterpillar

REPRODUCTION Eggs are laid singly on leaves of plants in the nightshade family. Larvae feed from the underside of leaves, where they rest on the midvein. Pupation takes place in a hollowed-out chamber several inches down in the soil. One generation is produced throughout the northern part of their range, where they fly all summer. From Virginia and North Carolina south and west to Texas, two generations may occur annually.

SIMILAR SPECIES The Carolina sphinx (*M. sexta*) has white and gray wing fringes and usually six pairs of abdominal spots. Its larva, the tobacco hornworm, has seven oblique lines along each side and a red-tipped horn.

Of Interest Tomato hornworms and tobacco hornworms are sometimes parasitized by the larvae of *Cotesia congregatus*, a small braconid wasp. The larvae chew their way out of the dying caterpillar to pupate inside egg-like cocoons.

WHY ARE MOTHS ATTRACTED TO LIGHTS?

Phototaxis is defined as an organism's movement in response to light. Cockroaches are negatively phototaxic and typically scurry away from lights, but many moths and other **nocturnal** insects are positively phototaxic and strongly attracted to lights. Entomologists have long exploited this behavior by using light traps to sample nocturnal insect populations. But what is the attraction?

DAZED AND CONFUSED

Several theories have been proposed to explain positive phototaxis in insects, especially in relation to their attraction to artificial lights. One prominent theory is that most moths and other nocturnal flying insects evolved using distant natural light sources such as the moon or stars to navigate. It was relatively easy to maintain a constant flight angle in relation to these distant heavenly bodies. Closer artificial sources of light, such as porch- and streetlights, require insects to continually adjust their flight angle as they pass by, resulting in a flight path that spirals directly toward the light source. Arriving at the light, the confused nocturnal moth goes into daylight mode and rests.

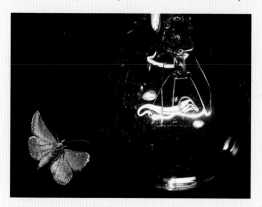

Moths are drawn to the artificial light of an electric lightbulb.

LEAVE THE LIGHTS ON AT HOME

Night-flying insects are attracted to all kinds of artificial lights, especially those with a strong ultraviolet component, such as the bluish mercury vapor lights. Depending on where you live, a floodlight aimed at a light-colored wall or white bed sheet suspended between two trees can be very productive for attracting all sorts of nocturnal insects. Warm, moonless nights are best for attracting them. The yellowish sodium vapor lights that are commonly used in many cities and towns are not particularly attractive to insects.

- Size: W 87.0–96.0 mm
- Family: **Sphingidae**
- Life cycle: **Probably one generation produced annually**
- Range: **United States, except Pacific Northwest and Great Basin**
- Food: **Adults sip nectar; larvae eat grape and Virginia creeper leaves**

Achemon Sphinx
Eumorpha achemon

*Although usually **nocturnal** and often attracted to light, the Achemon sphinx also visits flowers in the late afternoon and at dusk.*

IDENTIFICATION This large, robust moth is mostly tan brown with rusty-brown triangles on the **thorax.** The forewings have dark brown patches at the tips and middle, while the hind wings are mostly pinkish at their bases. Mature **caterpillars** are speckled green or brown with six three-part whitish or yellowish diagonal bands along the sides.

HABITS Adults visit Japanese honeysuckle, petunias, phlox, periwinkle, and evening primrose. They are also important pollinators of an endangered orchid in the upper Midwest.

REPRODUCTION After mating, females lay their eggs singly on the tendrils and undersides of leaves of grapes, Virginia creeper, and other members of the grape family growing along fencerows and woodland edges. The larvae dig chambers several inches underground before transforming into dull reddish brown. Feeding only at night, these horned caterpillars spend their days hiding among the leaves and branches of their food plant. They are occasionally reported to cause damage in vineyards and may become pests. More study is required in the South to determine whether the Achemon sphinx produces more than one generation annually or if the year-round presence of adults is due to an extended flight period.

SIMILAR SPECIES Of the 12 *Eumorpha* species that occur in North America, only *E. achemon* has hind wing bases that are mostly pink.

Of Interest When threatened or disturbed, the Achemon sphinx caterpillar can expand the last **thoracic** segment of its body and withdraw its head inside.

Caterpillar

- Size: W 63.0–90.0 mm
- Family: **Sphingidae**
- Life cycle: **One or more generations produced annually**
- Range: **Across southern Canada and United States**
- Food: **Adults sip nectar; larvae eat many kinds of plants**

White-Lined Sphinx
Hyles lineata

The white-lined sphinx— North America's most widespread and abundant hawk moth—occurs in gardens, fields, deserts, grasslands, and forest edges.

IDENTIFICATION The head and **thorax** are striped, while the olive-brown forewings have a broad tan stripe along their entire length and white veins. The hind wings are black with a pink band. Mature **caterpillars** have a horn on the tip of the **abdomen** and are variable in color, ranging from mostly black with yellow lines to mostly yellow with black lines, often with a yellow head.

HABITS In areas with dense populations, these mostly **nocturnal** moths are sometimes seen hovering over flowers. Population explosions of white-lined sphinxes sometimes occur in the West, where thousands of caterpillars are encountered crossing desert highways, and hundreds of adults swarm to lights at night.

Caterpillar

REPRODUCTION Mating takes place at dusk in spring and summer. The voracious caterpillars eat mostly plants in the evening primrose and rose families but will also devour many other kinds of plants. Northern populations produce one generation, while to the south they are capable of producing multiple generations. Adults are found year-round across the southern United States.

SIMILAR SPECIES The smaller and chunkier *H. galli* has brownish forewings without white veins and occurs in Alaska and Canada and across the northern United States. The European *H. euphorbiae*, introduced in the upper Midwest and Great Lakes region as a **biological control** agent of leafy and cypress spurges, resembles *H. galli*, but its forewings are more greenish.

Of Interest Because of their inclination to hover while sipping flower nectar, these are sometimes called "hummingbird moths."

- Size: **W 32.0–50.0 mm**
- Family: **Sphingidae**
- Life cycle: **One or more broods produced annually**
- Range: **Most of Canada and United States**
- Food: **Adults sip nectar from flowers; larvae eat leaves of honeysuckle and snowberry**

Snowberry Clearwing
Hemaris diffinis

Snowberry clearwings strongly resemble large bumble bees as they hover over flowers on warm days, but they have distinctly long **antennae** *and they never land.*

IDENTIFICATION Their bodies are clothed in varying degrees of brownish, yellowish, and black **setae.** Both pairs of wings are narrowly margined with dark **scales** and have large clear patches lacking scales except along the veins. Mature **caterpillars** have a long black horn on the tip of the **abdomen** and are either blue green down their backs and green on the sides or brown overall. Both color forms are covered with small white bumps and have a distinct black spot on each abdominal **spiracle.**

HABITS In spring and summer, adults visit flowers that they find growing in gardens, parks, and fields. They never

Caterpillar

land on flowers, but they extend their front legs and long **proboscis** forward as they feed.

REPRODUCTION Eggs are laid singly on the tips of new leaves of food plants that these moths find growing along power line cuts, fencerows, and edges of other open habitats. The caterpillars pupate on the surface of the ground within a cocoon consctructed of tough silk and debris. The broad range of individual, seasonal, and geographic variation in snowberry clearwings requires additional study.

SIMILAR SPECIES The larger hummingbird clearwing *(H. thysbe)* is similar in habit and occurs throughout the eastern United States and all of Canada. The top of the **thorax** varies from greenish to yellow or brown, while the underside is pale-yellowish white and the transparent wings have reddish-brown margins. The abdomen is uniformly colored, often matching the thorax, or has two segments that are reddish brown.

- Size: W 37.0–75.0 mm
- Family: **Erebidae**
- Life cycle: **One or more generations produced annually**
- Range: **British Columbia to Newfoundland, south to Utah, Texas, and Georgia**
- Food: **Adults probably don't feed; larvae eat plants in sunflower family**

Parthenice Tiger Moth

Grammia parthenice

The parthenice tiger moth is the most widespread species of Grammia *in North America.*

IDENTIFICATION The forewings of this stout, furry, and brightly colored moth are variably marked with thick off-white bars surrounding angular black patches and range from mostly off white to mostly black. The hind wings and **abdomen** are orange pink to deep pink or scarlet, with black spots limited to the outer half of the wing. Individuals of southeastern populations tend to be larger than those found elsewhere in their range. The mature **caterpillar,** or woolly bear, is dark with pale spots and a narrow stripe down the back, and bristling with long orangish-brown spines.

HABITS Adults, especially males, are commonly attracted to lights near fields, wet meadows, and edges of wetlands and woodlands in summer and early fall. Parthenice tiger moths are most common in the Rockies and Great Plains, from British Columbia and Alberta south to Arizona and New Mexico.

REPRODUCTION Females lay their eggs among vegetation on the ground rather than attaching them to larval food plants. Caterpillars are general feeders on low-growing plants, and they seem to be especially fond of the leaves of dandelions, ironweed, thistles, and their weedy relatives.

SIMILAR SPECIES There are 37 species of *Grammia* in North America, many with similar bold wing patterns. *Apantesis* species are similar but smaller than most species of Grammia.

Of Interest American folklore holds that the woolly bear of the Isabella tiger moth, *Pyrrharctica isabella*, predicts the severity of approaching winter based on the length of its orange middle band.

Caterpillar

- Size: W 26.0–43.0 mm
- Family: **Limacodidae**
- Life cycle: **One generation produced annually**
- Range: **Eastern United States**
- Food: **Adults probably don't feed; larvae eat leaves of many kinds of plants**

Saddleback Caterpillar
Acharia stimulea

*The striking mature **caterpillar**, sluglike and boldly marked, is more likely to be seen than the nondescript brown moth it eventually becomes.*

IDENTIFICATION Adults have shiny dark brown forewings with blackish areas, paler hind wings, and long, densely scaled legs. Each forewing has a white dot near the base and one to three dots near the tip. The mature caterpillar is brown or charcoal black on the ends, lime green edged in white across the middle, with a central brown saddlelike oval ringed with white. Fingerlike knobs bristling with stinging spines project from the **thorax** and tip of the **abdomen.** The front of the body is mostly plain, while a pair of large yellowish false eyespots marks the other end.

HABITS Adults emerge in summer, earlier in southern Florida and Texas, and even in areas where the caterpillars are common, they are only infrequently attracted to lights. Males begin flying at twilight, while females take wing after dark.

REPRODUCTION Mating pairs remain in copula for as long as 24 hours. Irregular clutches of 30–50 transparent yellow eggs are laid at night on the upper leaf surfaces of many different species of host plants. The larvae feed out of sight, on the undersides of leaves, where they glide sluglike on semifluid silk over the surface. Stinging spines are incorporated into the cocoon to protect the pupa overwintering inside.

SIMILAR SPECIES The uniquely marked saddleback caterpillar cannot be confused with any other North American butterfly or moth caterpillar.

Of Interest The spines of this caterpillar inflict painful stings that cause redness and prolonged discomfort.

Moth

Wasps, Bees & Ants

Order Hymenoptera

Ants are **eusocial,** but most species of bees and wasps are solitary.
Adults and larvae have chewing mouthparts adapted for eating plant
and animal tissues. Development is by **holometaboly.**

- Size: **L 24.0–28.0 mm**
- Family: **Sphecidae**
- Life cycle: **One generation produced annually**
- Range: **Across southern Canada and United States**
- Food: **Adults sip nectar; larvae eat paralyzed spiders**

Black-and-Yellow Mud Dauber

Sceliphron caementarium

Solitary females build mud nests in protected places.

IDENTIFICATION Dark wings fold over black-and-yellow bodies, with mostly yellow legs and a threadlike waist, or **petiole.**

HABITS Not particularly aggressive, female black-and-yellow mud daubers may defend themselves with a mild sting. In late summer, both sexes visit gardens, parks, fields, and wooded areas to find food and mates.

REPRODUCTION Females carry mud from puddles to build a nest of up to 25 cylindrical cells that are often all plastered over with mud to form a smooth nest. Each cell is stuffed with six or seven paralyzed arachnids—usually crab spiders, orbweavers, and jumping spiders—and sealed with a thick plug of mud. Larvae develop rapidly and winter as grubs enclosed in a parchmentlike cocoon within their mud cells.

SIMILAR SPECIES Of the three species found in North America, only *S. caementarium* is widespread.

Of Interest Shiny blue mud dauber *(Chalybion californicum)* females carry water to abandoned Sceliphron nests to refashion them for their own use.

Blue mud dauber female

- Size: **L 30.0–50.0 mm**
- Family: **Crabronidae**
- Life cycle: **One generation produced annually**
- Range: **Widespread east of Rocky Mountains**
- Food: **Adults drink nectar and sap; larvae eat paralyzed cicadas**

Eastern Cicada Killer
Sphecius speciosus

Lone eastern cicada killers patrol low over the ground in gardens, parks, and woodland edges during the summer.

IDENTIFICATION They have 12 (female) or 13 (male) **antennomeres,** reddish-and-black **thoraxes,** amber wings, and thick orangish legs. The **abdomen** is mostly black or brown with three pairs of bright-yellow markings.

HABITS Adults drink nectar and sap for their own nutrition and live about two weeks. Adult males stake out areas several meters square at sites where receptive females are most likely to appear. Perching on rocks and low branches, they engage in vigorous head butting, grappling, and biting with rival males. Mating pairs fly off together with the smaller male hanging head downward.

REPRODUCTION Females dig burrows in sandy roads and embankments or along sidewalks to a depth of 4 feet. Burrows are marked at their entrances by coarse mounds and consist of numerous branches and cells. Females drag or fly cicadas up to three times their body weight back to the nest, carrying them belly-to-belly and provisioning each burrow with up to four stung and paralyzed cicadas. A single egg is deposited on the last cicada placed in the nest. Females determine the sex of their offspring and will lay female (fertilized) eggs in the nest with the most abundant provisions.

SIMILAR SPECIES Four species of cicada killers occur in North America. *Sphecius hogardii* is restricted to southern Florida, while the others *(S. convallis, S. grandis)* occur in the western United States.

Of Interest Female cicada killers seldom sting humans, despite their fierce appearance, but they are sometimes considered nuisances when they nest in yards and play areas. Males of this species cannot sting.

Adult female dragging cicada

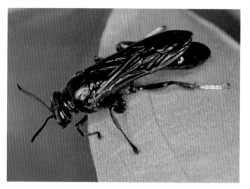

- Size: **L 13.0–19.0 mm**
- Family: **Crabronidae**
- Life cycle: **Two generations produced annually**
- Range: **Eastern United States**
- Food: **Adults sip nectar; larvae eat paralyzed spiders**

Pipe Organ Mud Dauber
Trypoxylon politum

Pipe organ mud dauber nests can be found on smooth, flat surfaces under protected overhangs such as eaves, chimneys, and bridges—even inside birdhouses.

IDENTIFICATION These wasps are nearly all shiny black, except for their pale hind feet. They have shiny dark wings with purplish-blue reflections.

HABITS They create nests of long, parallel mud tubes, up to 6 inches in length hence the name "pipe organ." Solitary females spend several hours constructing each tube, gathering mud and applying it in a series of interlocking arches. A male guards the nest, while the female hunts for spiders. Each nest tube is partitioned into several cells, and provisioned by the female with paralyzed orbweaver spiders.

Mud nest

REPRODUCTION Before egg laying begins, the female mates repeatedly with the male inside the nest. Females select the sex of their offspring by depositing female (fertilized) or male (unfertilized) eggs, and they provide the cells of their daughters with more food. Adults emerge through the sides of their individual cells in late spring and early summer.

SIMILAR SPECIES The genus *Trypoxylon* includes 29 species in North America: All construct mud nests and provision them with spiders, but only *T. politum* builds a mud nest consisting of parallel mud tubes. The other species utilize abandoned mud dauber nests, tunnels chewed by beetles and other wood-boring insects, crevices in wood and rocks, hollow stems, and wind chimes.

Of Interest Male pipe organ mud daubers will guard the nest entrance to drive off parasites while the female is away.

- Size: L worker 9.0–18.0 mm; drone 15.0–16.0 mm; queen 18.0–20.0 mm
- Family: Apidae
- Life cycle: Multiple broods produced annually
- Range: Across southern Canada and United States
- Food: Adults eat nectar, honey; larvae eat honey

European Honey Bee
Apis mellifera

European honey bees, most valued for their pollination services, were introduced to North America by British colonists.

IDENTIFICATION Workers are golden brown and black, with four clear wings and pale-yellow bands on the **abdomen.** Unlike **queens** and **drones,** workers have hind legs modified for collecting and carrying pollen.

HABITS European honey bees live in colonies with up to 80,000 individuals consisting mostly of workers, some drones, and a single queen. Wild colonies establish hives in hollow trees, beneath rock overhangs, in wall voids, and under bridges. Workers care for the queen and drones, raise the brood, forage for food, and defend the colony. They defend the hive by using their modified **ovipositors** as stingers, leaving behind not only the barbed stinger in the flesh of their

Queen and drone in hive

attacker but also some of their vital internal organs, and so they die soon afterward. Queens can sting multiple times but use their stingers only to kill rival queens. The sole purpose of drones is to mate with future queens. Unable to care for themselves, they are expelled from the hive in autumn. European honey bees are also highly prized for their honey, wax, pollen, bee venom, and other products. Habitat destruction, pesticides, **parasites,** and changing weather patterns threaten their ability to pollinate food crops for humans and domestic animals.

REPRODUCTION Queens mate only once, live up to five years, and lay up to a million eggs in a lifetime. Workers and queens develop from fertilized eggs, drones from unfertilized eggs.

SIMILAR SPECIES Drone flies (*Eristalis sp.*; p. 147) resemble European honey bees and are often found nectaring on flowers. They have only two wings and large eyes resembling those of drones, and they are incapable of biting or stinging.

- Size: **L worker 11.0–17.0 mm; drone 13.0–16.0 mm; queen 18.0–21.0 mm**
- Family: **Apidae**
- Life cycle: **Multiple broods produced annually**
- Range: **Across southern Canada and United States**
- Food: **Adults sip nectar, honey; larvae eat pollen and honey**

Yellow Bumble Bee
Bombus fervidus

Yellow bumble bees, also called golden northern bumble bees, live in colonies of 50 to 125, led by a **queen.**

IDENTIFICATION The hairy workers have black heads and an entirely yellow **thorax** that may have a black band between the bases of the smoky brown wings. The **abdomen** is usually mostly yellow with a black tip, but it may be mostly black. This widespread species tends to exhibit darker color patterns in the West. They have relatively long mouthparts and are able to collect nectar from deep-throated flowers.

HABITS They occupy various habitats, including grassy areas in gardens, parks, meadows, roadsides, and forest openings. Workers have smooth stingers and will use them repeatedly in defense of themselves and the nest. Colonies persist only for a single season. Like other native bees, this and other bumble bees are important pollinators, but populations are declining owing to habitat loss and disease.

REPRODUCTION Virgin queens and **drones** develop in the fall and mate. Mated queens overwinter and emerge in spring to establish new colonies. They build nests of grass above ground, in abandoned rodent burrows, or in other sheltered spaces underground. The nest floor contains potlike cells of wax containing honey for the brood.

SIMILAR SPECIES Forty-six species of bumble bees occur in Canada and the United States. Many species exhibit variable color patterns across their ranges and are difficult to identify by sight alone.

Of Interest Bumble bees keep warm on cool days by disengaging their wings and shivering their flight muscles, thus raising their thoracic temperature to allow flight.

Bees in nest

BRINGING POLLINATORS HOME

Insects are responsible for pollinating 85 percent of all flowering plants and about two-thirds of the world's crop species. Most terrestrial ecosystems depend on bees, flies, moths, and beetles to pollinate flowers and maintain plant and animal populations.

Natural habitats keep shrinking, and most modern landscapes are sterile shrines celebrating our obsession with keeping things neat and tidy. Carpeted with incessantly mown grass and devoid of decaying leaves, fallen branches, and crumbling logs and stumps, many neighborhoods are inhospitable wastelands for pollinators and other wildlife. Native bees and other insects require bare patches of ground and holes, nooks, and crannies in wood and stone in which to build nests, brood their young, and undergo metamorphosis.

Following are some ideas on how to bring the pollinators back home.

Insect hotels made from natural and manufactured items provide nesting sites and shelter for all kinds of insects, spiders, and other arthropods.

By planting shrubs and flowers that attract pollinators, you can bring interesting arthropods close to home and benefit other wildlife.

GO NATIVE

Provide pollen and nectar for pollinators and other wildlife by cultivating species of native plants appropriate for your region. Select a diversity of annuals and perennials, including woody shrubs and trees. Diversity in plants and their structure supports a greater diversity of pollinators and other wildlife. Consult local native plant societies and government agencies for suggestions of suitable native plants.

CREATE NESTING SITES

Make nesting blocks using lumber free of preservatives and drilled with smooth nesting holes of various diameters (3/32 inches and 3/8 inches) and depths (3–6 inches). Leave downed wood and stumps in place, or add some in sunny locations, drilling holes similar to those of the nest boxes on their southeastern exposures. Cut bamboo and other hollow reeds into 6- to 8-inch lengths and bundle them together. Or create an insect hotel that provides lots of nesting holes with a combination of natural and manufactured materials. Nests should be attached securely in a site that remains dry with their entrances facing east or southeast to catch the morning sun.

STOP USING PESTICIDES

Along with loss of habitat and competition from invasive species, pesticide use is a leading cause of population decline in native insect pollinators and other wildlife. Do not use pesticides, herbicides, or fungicides in your native plant garden, and encourage your neighbors to do the same. Remember that healthy landscapes with sufficient food and habitat will support native species of **predators** and **parasites** that will control insect pests. Using pesticides will only harm these beneficial insects and exacerbate pest problems.

Visit xerces.org for information on native bees and other pollinators.

- Size: L 20.0–25.0 mm
- Family: **Apidae**
- Life cycle: **One generation produced annually**
- Range: **New England to central Florida, west to Nebraska and Texas**
- Food: **Adults sip nectar; larvae eat pollen**

Eastern Carpenter Bee
Xylocopa virginica

Overwintering eastern carpenter bees emerge from their nest chambers and other protected places in spring, filling the warm air with their noisy buzzes.

IDENTIFICATION These are large, robust, shiny blue-black bees with a head nearly as wide as the **thorax** and very short or no hairlike **setae** on the **abdomen.**

HABITS Dark-faced females are capable of delivering a defensive sting, but they seldom do so. White-faced males cannot sting, but they will aggressively drive off rival males and intimidate humans who venture into their ever-changing territories.

REPRODUCTION Courting pairs engage in noisy aerial acrobatics, flying apart and coming together several times over. For nests, solitary females chew tunnels in dead wood and exposed structural timbers, railings, furniture, and other

wooden structures, preferring sites with eastern or southern exposures. The single shallow entrance is perfectly round, about half an inch in diameter, and extends only about a body length into the wood. One or more tunnels up to an inch long are chewed at right angles to the entrance, paralleling the wood's grain. Inside a tunnel, up to eight cells each contain a single egg and a doughy pill of pollen about the thickness of a kidney bean. Bees emerge briefly in late summer before hiding away for the winter.

SIMILAR SPECIES Large bumble bees resemble eastern carpenter bees, but they have relatively smaller heads and one or more distinct bands of hairlike setae on their abdomens.

Of Interest As they forage for pollen and nectar, carpenter bees mark each flower visited with a **pheromone** so others know to avoid blooms recently depleted of nectar.

Close-up of male head

- Size: L 15.0–25.0 mm
- Family: **Mutillidae**
- Life cycle: **One generation produced annually**
- Range: **Eastern United States**
- Food: **Adults drink nectar; larvae eat bumble bee brood**

Cow Killer

Dasymutilla occidentalis

The largest and most widely recognized velvet ant in North America, the cow killer is actually a solitary female wasp. Despite its name, this species does not kill cows.

IDENTIFICATION Hairy and antlike, cow killers are bright red with bold black markings. Males have four dark wings, and the front half of the **abdomen** is black, while the wingless females have a black band across the middle of the abdomen. The red-and-black coloration is **aposematic** and serves as a warning to potential **predators.**

HABITS Females defend themselves by inflicting a sting that causes excruciating pain. Stingless males benefit from their similarly bold and contrasting pattern, which misleads their enemies into thinking that they, too, can sting. Males are often seen during the summer flying low over the ground in open habitats and along woodland edges as they search for females scurrying on the ground. Females may live more than 500 days in captivity, suggesting that they might overwinter as adults in the wild and persist through almost an entire second season.

REPRODUCTION After mating, the females enter the subterranean nests of bumble bees and, occasionally, great golden digger wasps to lay their eggs. After hatching, the developing larvae consume the grubs and pupae.

SIMILAR SPECIES *Dasymutilla* species are sometimes difficult to identify. Most occur in the West, including the mostly white *D. gloriosa* and *D. sackenii*, the black-and-red *D. klugii* and *D. magnifica*, and the mostly red *D. aureola*.

Of Interest Both sexes **stridulate,** using their abdominal segments to produce squeaking sounds during mating and when disturbed.

Adult male

- Size: L 14.0–50.0 mm
- Family: Pompilidae
- Life cycle: Southern half of United States
- Range: One generation produced annually
- Food: Adults drink nectar and sap; larvae eat tarantulas and other large spiders

Tarantula Hawks

Pepsis and *Hemipepsis* species

Large, conspicuously iridescent wasps borne on orange wings, tarantula hawks are familiar sights in the West as they drink nectar from milkweeds and other flowers on hot summer days.

IDENTIFICATION These robust wasps are black, sometimes with distinct bluish or greenish reflections. The **antennae** are typically straight (male) or curled (female). Their wings are dark to bright reddish orange. The bright colors of these solitary wasps are **aposematic,** warning **predators** of the female's painful sting. As with all other ants, bees, and wasps, the males are incapable of stinging but are protected by their similarity to females. The incredible pain generated by the female's sting is likely the evolutionary basis for similarly colored yet unrelated and harmless insects **(Batesian mimicry)** such as the longhorn beetle, *Tragidion*, and a large fly, *Mydas luteipennis.*

Wasp carrying tarantula

HABITS Tarantula hawks are especially abundant during the summer in the arid regions of southern California and the Southwest. The hard, smooth body and quick reflexes of the females give them an advantage over much larger tarantulas, which they paralyze with their stings and stuff down into burrows as food for their developing wasp grubs.

REPRODUCTION Fertilized eggs develop into females, unfertilized eggs into males.

SIMILAR SPECIES There are 19 species of *Pepsis* and three species of *Hemipepsis* found north of Mexico. Species of *Hemipepsis* are black and generally lack the distinct bluish or greenish reflections found in *Pepsis* species.

Of Interest The pain may persist for up to a week, but the tarantula hawk's sting is generally not considered dangerous enough to warrant medical attention.

- Size: L 9.0–19.0 mm
- Family: Vespidae
- Life cycle: Two generations produced annually
- Range: Southern Ontario and eastern United States
- Food: Adults drink nectar; larvae eat caterpillars

Potter Wasp
Eumenes fraternus

Female potter wasps build marble-size mud nests with funnel-shaped spouts, each housing a single egg and provisioned with paralyzed moth **caterpillars.**

IDENTIFICATION These solitary wasps are black with yellowish-white markings. They have distinctly brownish forewings with violet reflections and a slender, cone-shaped waist that is narrow in front and as long as the **thorax.** Males are more slender than females, have pale faces, and **antennae** tipped with clawlike hooks.

HABITS Adults emerge in late spring and summer to sip nectar from flowers and seek mates. Females attach their nests singly or in groups of two to five on rocks, stems, and tree trunks in open habitats and woodlands, or to structures made of wood. They construct the nests by mixing water with dry soil and fashion the mud into pellets. One by one, they carry the pellets to the nest site with their **mandibles** and forelegs. Although it takes hundreds of mud pellets, the female can complete a nest in two hours or less.

REPRODUCTION Inserting the tip of her **abdomen** through an opening in the nest's narrow neck, the female lays a single egg inside, suspended on a filament. Then she flies off in search of pestiferous spring cankerworms and parsnip webworms, food for the developing larva. She stuffs their paralyzed bodies through the funnel-shaped spout and seals the opening with mud. Adults emerge from the sides of the pots.

SIMILAR SPECIES Eight species of *Eumenes* occur in Canada and the United States.

Of Interest Potter wasps make surprisingly good garden companions as they hunt many species of leaf-eating caterpillars that are considered pests.

Mud nest

- Size: **L 16.0–19.0 mm**
- Family: **Vespidae**
- Life cycle: **Several broods produced annually**
- Range: **Ontario; New Jersey and Florida west to California**
- Food: **Adults eat sugary fluids; larvae eat insects**

Paper Wasp
Polistes exclamans

The paper wasp Polistes exclamans is one of the most widespread and intensely studied social wasps in North America.

IDENTIFICATION Workers, males, and **queens** have reddish, yellow, and black markings on the head, **thorax,** and **abdomen.** They have long legs and a distinct yellow spot on each side of the thorax. Males have more black than red on the thorax and abdomen. The **antennae** have diffuse bands of yellow, black, and red.

HABITS Paper wasps build open paper nests in sheltered places on tree branches or beneath the eaves of homes, constructing them from fibers gathered from plant stems, posts, and fences. They mix the fibers with their saliva and fashion the resulting material into a single layer of hexagonal cells that open downward: nurseries in which to

brood the young. Nests are small, seldom housing more than 100 wasps, and persist only one season. Workers feed the larvae chewed-up insects but also take in nectar, **honeydew,** plant sap, fruit juices, and other sweet fluids for their own sustenance. When the larval occupant is ready to pupate, the cells are covered.

REPRODUCTION Each nest has one egg-laying queen, but the workers are physiologically capable of laying eggs should the queen die—in which case the oldest worker in the nest usually becomes the new queen.

SIMILAR SPECIES There are 19 species of *Polistes* in Canada and the United States.

Of Interest Paper wasp nests are often destroyed by birds or attacked by predatory ants or parasitic wasps, especially early in the season when the founding queen is the sole adult occupant of the nest.

Wasp on nest

- Size: L workers 18.0–35.0 mm; queen 20.0–38.0 mm
- Family: Vespidae
- Life cycle: Multiple broods produced in a season
- Range: Europe; established in southern Canada and eastern United States
- Food: Adults drink nectar, sap, and honeydew; larvae eat insects

European Hornet

Vespa crabro

The European hornet is the largest social wasp in Europe and North America.

IDENTIFICATION This insect's head and **thorax** are mostly reddish brown, while the **abdomen** is mostly black in front and yellow in the rear. The amber wings are narrow, and the legs are reddish brown.

HABITS European hornets build their enclosed paper nests in tree hollows in gardens, parks, and woodlands, but they will sometimes establish them under porches and overhangs, or in outbuildings and basements, becoming a nuisance. At the height of the season, up to a thousand individuals may occupy a single nest. Active day and night, workers are sometimes attracted to lights. The workers and the **queen** are capable of delivering a painful sting in their own defense.

Colonies persist for only one season, each dominated by a fertile queen and sterile female workers. Though sometimes considered pests when they strip bark off young branches from apple trees and lilac bushes to build their nests, these wasps are significant **predators** of garden insect pests and help control their populations. Although European hornets are capable of capturing insects as large as cicadas and dragonflies, up to 80 percent of their diet consists of flies.

REPRODUCTION Reproductive males and females appear in late summer and fall. Insect prey is chewed into a pulp and fed to larvae suspended upside down in layers of honeycombed brood cells stacked in the nest.

Brood cell layers

SIMILAR SPECIES European hornets are sometimes confused with the large and solitary eastern cicada killer, *Sphecius speciosus* (p. 184), which has a relatively small, dark head and an abdomen that is mostly black with pairs of broad yellow spots.

- Size: L worker 11.0–15.0 mm; queen 17.0–20.0 mm
- Family: Vespidae
- Life cycle: Multiple broods produced annually
- Range: Western North America
- Food: Adults eat nectar, plant sap, and sweet fluids; larvae eat invertebrates

Western Yellowjacket
Vespula pensylvanica

The most commonly encountered yellowjacket in the West, these insects frequently establish their nests in yards, parks, campgrounds, and outdoor recreation areas.

IDENTIFICATION These robust black-and-yellow wasps have a black diamond-shaped mark on top of the first **gaster** segment and a continuous broad yellow band or loop above each eye; males occasionally lack the eye loop.

HABITS Their gray, scalloped paper **carton** nests resemble those of hornets but are usually built in cavities in the ground, especially in abandoned rodent burrows near well-watered yards. They occasionally establish their nest in an attic, wall void, or other aboveground site. Inside the multilayered cover is a stack of rounded combs composed of hexagonal brood cells attached one below the other. Most nests contain 3,000 to 5,000 wasps. In warmer climates, these nests can become huge and persist for more than one year. Population outbreaks, especially in the mountains, occur every three to five years or more and are associated with warm, dry spring weather, when they become pests by scavenging flesh and sweets at picnics, outdoor dining areas, and garbage cans. Stings sometimes result in painful swelling, followed by a period of persistent itching.

REPRODUCTION Workers prey on insects, other arthropods, and slugs to chew up and feed to the larvae.

SIMILAR SPECIES Thirteen species of *Vespula* occur in North America, including the **adventive** German yellowjacket, *V. germanica*. The eastern yellowjacket (*V. maculifrons*) is the most common yellowjacket in eastern North America and has a distinctive anchor- or arrowhead-shaped black mark on top of the first gaster segment.

Brood cell layers in underground nest

- Size: L worker 2.0–3.0 mm; queen 4.2–6.4 mm
- Family: **Formicidae**
- Life cycle: **Multiple broods produced annually**
- Range: **Western and southeastern United States**
- Food: **Honeydew and dead insects**

Argentine Ant
Linepithema humile

Originally from South America, Argentine ants are among the world's most successful invasive species, inadvertently distributed by humans to six continents and many oceanic islands.

IDENTIFICATION These tiny ants are uniformly light to dark brown, have **mandibles** with small **denticles** interspersed between larger teeth, and lack erect **setae** on the first two segments of the **gaster.** There are 10 **antennomeres,** the last two of which form a club. The narrow waistlike **petiole** has a single, flattened **node.**

HABITS They live in urban and suburban areas, agricultural fields, and other disturbed areas and do not sting or bite. Nests are established in small spaces, including cracks in walls and sidewalks, gaps in walls and potted plants, and within the nooks and crannies afforded by compost, leaf litter, wood chips, mulch, and debris. They can establish very large colonies and are frequently household pests.

REPRODUCTION Workers tend aphids and other sap-sucking insects to obtain **honeydew** for the larvae; they also protect them from **predators** and **parasitoids.** Their presence often hampers **biological control** efforts in nurseries, croplands, and orchards, especially in citrus groves. Individuals from different nests seldom attack one another and may be part of giant supercolonies, one of which stretches 560 miles along the California coast.

SIMILAR SPECIES Argentine ants are distinguished from other ants by their antennae and petioles.

Of Interest Argentine ants frequently displace native ant species. They are blamed for the decline of the coast horned lizards in California, which depend on red harvester ants for food.

Adult

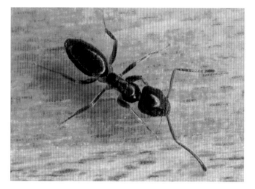

- Size: L worker 2.4–3.3 mm; queen 3.7–4.9 mm
- Family: **Formicidae**
- Life cycle: **Multiple broods produced annually**
- Range: **Across southern Canada and United States**
- Food: **Honeydew and foodstuffs containing sugar; dead insects and spiders**

Odorous House Ant
Tapinoma sessile

The native odorous house ant is one of the most widespread ants in North America and is so named because of the unpleasant odor it releases when crushed.

IDENTIFICATION These ants are small, uniformly dark brown or black with lighter mouthparts and legs. The **antennae** have 12 **antennomeres** and lack a distinct club. The slender, waistlike **petiole** lacks distinct **nodes** and is hidden from above by the **abdomen.**

HABITS Odorous house ants live in colonies with more than 10,000 workers and multiple **queens** occupying several nests. Ants nest indoors in homes, apartments, and greenhouses under appliances, in walls, and under mats. Colonies typically overwinter in a single nest but will disperse to nearby spaces in walls and other cavities when food becomes more plentiful in spring and summer.

Outside they will live under rocks, inside or beneath potted plants, and even inside patio furniture and cars. Workers forage day and night for **honeydew** produced by aphids, treehoppers, and scale insects. They usually begin moving indoors in spring to search for sugary foods and water. They do not sting or cause structural damage. Odorous house ants are best controlled indoors by tidying up food scraps and removing open sources of water. Outdoors, keep mulch and potted plants away from the house.

REPRODUCTION Winged males and females emerge in summer to mate.

SIMILAR SPECIES Odorous house ants are distinguished from other small, uniformly brown or black ants by the lack of nodes on the petiole.

Of Interest Recent genetic studies suggest that this ant may be a **complex** of several closely related species.

Petiole hidden by abdomen

SUPERCOLONIES

Ants live in groups, or colonies, that cooperate in rearing young, gathering food, and defending the nest. Depending on species, ant colonies may contain a few dozen to millions of ants. Individuals within these colonies typically recognize one another by smell and will ferociously attack ants from other colonies, even if they are of the same species. Super-colonies exceed one million individuals and grow without restriction under the right conditions. They form when **queens** and workers move together to new sites **(budding),** thus increasing the number of nests within a single colony, or by engaging in long-distance movements **(jump dispersal),** usually as a result of human activities.

GOING GLOBAL

Invasive colonies of Argentine ants *(Linepithema humile)* con-sisting of closely related individuals form giant supercolonies containing billions of individuals. These supercolonies are the largest animal societies known and are matched only by human civilization. In fact, in 2009 it was demonstrated that the Argen-tine ant supercolonies in Europe, California, and Japan were very closely related to one another, as determined by the chemical odors of their **exoskeletons** and their nonaggressive and grooming behaviors when interacting with one another—leading researchers to conclude that these three colonies actually represent a single global supercolony.

Some of the world's largest ant super-colonies can stretch across entire states, countries, and continents.

- Size: L worker 5.0–16.0 mm; queen 19.0 mm
- Family: Formicidae
- Life cycle: One brood of workers produced annually
- Range: Eastern North America
- Food: Insects, honeydew, and plant sap

Black Carpenter Ant

Camponotus pennsylvanicus

The black carpenter ant is one of the largest ants in North America.

IDENTIFICATION These ants, black at a glance, are sparsely covered with golden hairlike setae on the head, **thorax**, and **gaster**. The slender, waistlike **petiole** has one **node.**

HABITS Black carpenter ants nest in living deciduous trees and decaying logs found in urban and suburban areas, parks, and woodlands. Colonies can include up to 2,000 individuals; a colony consists of both smaller (minor) ants and larger (major) workers. They chew their galleries and chambers parallel with the grain of the wood, using powerful **mandibles** and hasten the wood's decay in the process. They will attack structural timbers, but only when the wood is already compromised by moisture or termites and other insect pests. Unlike termites, these ants do not eat wood.

Workers forage for food mostly at night and may wander long distances in search of insect prey and sweet fluids.

REPRODUCTION After several years, colonies produce winged males and females that engage in nuptial flights at the height of summer. As with all other ants, the male ant dies shortly after mating, while the newly mated **queen** sheds wings and seeks a sheltered spot to establish a new colony. Workers may live up to 7 years, while queens 10 or more years.

SIMILAR SPECIES There are about 50 species north of Mexico. The western carpenter ant *(Camponotus modoc)* occurs in western North America and is similar but has reddish legs.

Of Interest These ants cannot sting, but when disturbed they are capable of inflicting a painful bite that is exacerbated by their spraying formic acid from the tip of the **abdomen** onto the wound.

Winged queen

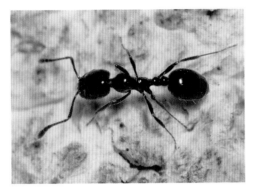

- Size: L worker 1.0–2.0 mm; queen 4.0–5.0 mm
- Family: **Formicidae**
- Life cycle: **Multiple generations produced annually**
- Range: **Throughout much of North America**
- Food: **Honeydew, nectar, pollen, and insects**

Little Black Ant
Monomorium minimum

Little black ants often invade homes in search of sweets and greasy foods. They can be seen traveling in conspicuous columns across kitchen walls, counters, cabinets, and floors.

IDENTIFICATION They are small, shiny reddish brown to black, with 12 **antennomeres,** the last three of which form a club. The slender, waistlike **petiole** has two **nodes.**

HABITS Highly adaptive, they build their nests in various disturbed habitats out in the open or under objects, including lawns, beneath rocks, or in decayed wood. They will also establish nests in woodwork and masonry. Their nest entrances may or may not be surrounded by small craters of fine soil. Colonies consist of a few thousand workers (just one form of worker; no soldiers) and many **queens.** For most of the year, the colony's inhabitants consist primarily of workers and the brood. Workers scavenge for dead insects and other arthropods and use **pheromones** in order to signal to their nest mates about food sources. They defend their booty from other ant scavengers by "**gaster-**flagging": in other words, by waving their **abdomens** in the air while simultaneously releasing repellent venom from their stingers—although they are too small to inflict painful stings on humans. Little black ants also tend aphids to obtain **honeydew.**

REPRODUCTION Winged males and females mate in summer.

SIMILAR SPECIES Widespread pharaoh ants (*M. pharaonis*) nest indoors and out. They are similar in size but are rough and pale with a dark-tipped abdomen. *Monomorium* species are often mistaken for fire ants (*Solenopsis*) but have antennal clubs with three, not two, antennomeres.

Nest entrance

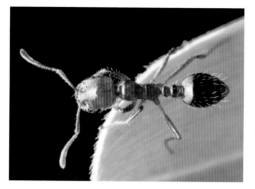

- Size: L worker 2.4–6.0 mm; queen 8.0 mm
- Family: Formicidae
- Life cycle: Multiple broods produced annually
- Range: Virginia to Florida, west to southern California
- Food: Plant tissue, honeydew, insects, and small vertebrates

Red Imported Fire Ant
Solenopsis invicta

First reported in Alabama in the 1930s, these fire ants are not only a nuisance but also serious agricultural pests causing major environmental damage.

IDENTIFICATION Smooth, shiny, reddish brown to black, they have 10 **antennomeres,** the last two forming a club. The stemlike **petiole** between the **thorax** and **abdomen** has two **nodes.**

HABITS Originally a native of Argentina, this species has spread throughout the southeastern United States, across the Southwest to southern California, and much of South America, the Caribbean, Australia, and parts of Asia. These ants live in colonies containing millions of smaller (minor) and larger (major) workers and one or more **queens.** Their nest mounds can reach 18 inches in diameter and height, so big they can hamper farm equipment that comes upon them. Colonies are accidentally

moved in shipments of turf, nursery plant root balls, and other horticultural products. Masses of ants also float on floodwaters. They can have an impact on infrastructure: Thousands of individuals will pack themselves into electrical boxes, shorting out traffic lights and air conditioners. They also cause environmental damage and cost billions of dollars annually, damaging crops and livestock. Efforts to control them severely impact native ants and other wildlife. Red imported fire ants bite before delivering a venomous sting that produces a burning sensation followed by the development of a red, fluid-filled pustule. Multiple stings can seriously injure or kill caged animals and nestlings.

REPRODUCTION Queens produce as many as 200 eggs per day.

SIMILAR SPECIES There are 40 mostly native *Solenopsis* species in the United States. The South American black imported fire ant (*S. richteri*) and the native southern fire ant (*S. xyloni*) occur in the Southeast.

Nest mound

Spiders, Ticks & Allies

Class Arachnida

Arachnids have two major body regions and eight legs as adults. They have pincherlike mouthparts, or **chelicerae**, modified into fangs among spiders. They are **predators,** scavengers, or plant feeders.

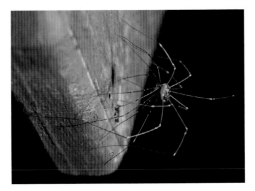

- Size: L 6.0–9.0 mm
- Family: **Pholcidae**
- Life cycle: **One generation produced annually**
- Range: **North America**
- Food: **Insects and other spiders**

Long-Bodied Cellar Spider

Pholcus phalangioides

This is the most common cellar spider in North America.

IDENTIFICATION These pale gray spiders have eight eyes, long legs, and an **abdomen** twice as long as wide. The round, flat **carapace** is sometimes dark around the eyes and down the middle; the abdomen may have dark markings.

HABITS Most common around homes and buildings in the North but also outdoors in warmer southern regions, these spiders construct webs indoors in dark, undisturbed places, such as ceiling corners or under bookshelves. They draw silk from their **spinnerets** to entangle approaching insect and spider prey, even those larger than themselves. When threatened, they vibrate back and forth in their webs, apparently to make themselves hard to see. Their venom is no stronger than that of a honey bee, and their weak fangs are incapable of puncturing human skin.

REPRODUCTION Females wrap their eggs in a thin silken sac and carry it in their mouthparts until hatching. Young spiders mature in a year, and adults may live two or more years.

SIMILAR SPECIES They are sometimes confused with harvestman or daddy long-legs, which do not produce silk or venom. An Old World marbled cellar spider, *Holocnemus pluchei*, is now established throughout the United States, especially in the Southwest.

Female guarding egg sac

- Size: **L 3.8–7.0 mm**
- Family: **Theridiidae**
- Life cycle: **One generation produced annually**
- Range: **Across southern Canada and United States**
- Food: **Insects**

Common House Spider
Parasteatoda tepidariorum

The common house spider is one of the most widely distributed spiders in the world and is found coast to coast in North America.

IDENTIFICATION The female common house spider is gray or tan with a yellowish brown or tan **carapace**. The **abdomen** varies in color from whitish to brownish and has a series of indistinct and grayish chevrons on top, near the **spinnerets;** the abdomen appears teardrop-shaped when viewed from the side as the spider hangs upside down in the web. The yellowish legs are banded near the end of each segment. Males are smaller, lighter in color, and somewhat reddish or orangish.

HABITS Common house spiders are found year-round in homes, basements, barns, and other outbuildings, especially in the North. Farther south they are often found outdoors under debris, on fences, and beneath bridges. Females create messy tangled cobwebs, often in windows and corners, where they hang upside down awaiting prey. Alongside them may be found the remains of the past insect meals, wrapped mummylike in silk; older prey items are eventually dropped from the web.

REPRODUCTION Females produce up to four brown, pear-shaped egg sacs wrapped in tough, parchmentlike silk. When threatened, they will hide behind debris or drop from the web on a silk line. Upon hatching, the young spiders will briefly remain in a tight cluster before dispersing. Adults may live a year or more after reaching maturity.

Egg sac in web

SIMILAR SPECIES A similar species, *Cryptachea rupicola,* has distinctly banded legs and a small black-and-white bump at the rear of the abdomen. Its cobweb includes a silken retreat festooned with bits of leaves and debris.

- Size: L male 3.0–6.0 mm; female 8.0–10.0 mm
- Family: Theridiidae
- Life cycle: One generation produced annually
- Range: Eastern United States, especially southern states
- Food: Crawling insects, spiders, and other arthropods

Southern Black Widow

Latrodectus mactans

Southern black widows are shy and seldom bite, but their venom contains a neurotoxin harmful to humans—so any bite requires immediate medical attention.

IDENTIFICATION Mature, shiny black females have a distinct and usually complete reddish hourglass marking underneath the nearly spherical **abdomen,** the lower portion of which is more rectangular then triangular. There is a row of red spots down the middle of the **dorsal** side of the abdomen. Males and immatures are both marked with white lines along the sides of the abdomen.

HABITS Females build messy cobwebs of strong, sticky silk low to the ground along rock walls, in woodpiles, or inside meter boxes and the entrances of animal burrows. Silk runners to the ground entangle crawling insects and other arthropods that are hauled up into the web, killed, and eaten. The much smaller male lurks in or near the female's web and is occasionally consumed by the female after mating. Females may live up to three years or more.

REPRODUCTION Females wrap their eggs in a whitish or beige parchmentlike silk sac, suspend it in the web, and guard the eggs until they hatch.

SIMILAR SPECIES The female northern widow, *L. variolus,* is a slightly larger spider with the hourglass broken into two separate parts. This species produces a more pear-shaped egg sac. It occurs in eastern North America, more commonly in the North. The female western widow, *L. hesperus,* typically has a complete hourglass pattern composed of triangular halves and occurs in western North America.

Of Interest In this species' scientific name, *Latrodectus* is Greek for "biting in secret," which refers to the initial painlessness of the black widow's rare bite.

Adult male

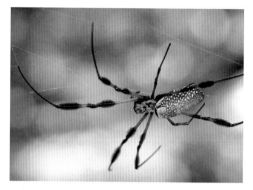

- Size: L male 4.0–8.0 mm; female 22.0–25.0 mm
- Family: Tetragnathidae
- Life cycle: One generation produced annually
- Range: Southern Atlantic and Gulf Coasts; South Carolina to Florida, west to Texas
- Food: Flying insects

Golden Silk Orbweaver
Nephila clavipes

Golden silk orbweavers, also called calico or banana spiders, live mostly in warm, humid habitats, especially near the edges of swamps and shaded hardwood and palm forests.

IDENTIFICATION The female's **cephalothorax** is densely covered with short, silvery hairlike **setae,** while the long **abdomen** is yellowish, orangish, or greenish with five or six pairs of white spots. All but the third pair of legs have conspicuous tufts of setae. The much smaller male is dark brown overall and lacks leg tufts.

HABITS Females construct large and slightly tilted webs that stretch up to 3 feet across the lower branches of trees and trails. Freshly produced silk is sticky and yellow but soon fades to white as it dries out. Spokes radiating outward from the web's center appear notched, rather than straight as in the webs of other orbweaver spiders. During the day, females hang head-down on the lower side of the web and wait for flying insects to become entangled, while the tiny males are usually found at the margins of the web or sometimes clinging to the underside of the female's abdomen. Small and silvery *Argyrodes* spiders often inhabit the periphery of their webs and will steal their prey. Large numbers of these **kleptoparasitic** spiders may drive *Nephila* spiders to relocate their webs.

SIMILAR SPECIES Golden silk orbweavers cannot be confused with any other North American spider.

Of Interest Engineers look to spider silk to design new materials. Per unit weight, the tensile strength of this spider's silk is much greater than that of steel.

Adult female (left) and adult male

SPIDER BITES ARE RARE

Spiders are needlessly maligned animals. Of the few spider bites that do occur, most are of little concern. Only on rare occasion does **envenomation** by a spider bite cause serious health issues or death.

BITES VERSUS STINGS

Arthropods can bite or sting. They use **mandibles** or fangs to bite; they use structures at the other end of the body to sting. Honey bees and some ants can do both, but their bites are inconsequential when compared with their stings. Spiders bite with their fangs.

As its name suggests, the brown recluse spider tends to be very shy and will often run away or even play dead when threatened.

POISONOUS OR VENOMOUS?

Poisons are exuded from glands or other body parts and are delivered topically or through ingestion. Venoms are also produced in glands, but they are delivered through ducts into hardened stingers, fangs, or spines, which inject them into the skin.

WIDOWS AND RECLUSES

There are only two groups of medically important spiders in North America: widows (*Latrodectus sp.*) and recluses (*Loxosceles*). Widow bites contain neurotoxins that result in moderate to severe systemic reactions involving the muscle and nervous systems with little injury to the skin. Recluse bites cause mostly skin damage, rarely with any systemic reactions, and are mostly minor and self-healing.

MRSA AND OTHER BACTERIAL INFECTIONS

Recent studies reveal that many skin lesions attributed by doctors and patients to spiders, especially those of *Loxosceles*, were actually MRSA and other bacterial infections. Although it is feasible that spider bites could be the point of introduction for bacterial infections, evidence for this is lacking. In fact, spider venom is well known for its antibacterial properties and its use in treating infections resistant to antibiotics is currently under investigation.

■ Size: L male 5.0–8.0 mm; female 19.0–28.0 mm
■ Family: Araneidae
■ Life cycle: One generation produced annually
■ Range: Southern Canada and throughout United States
■ Food: Flying insects

Black and Yellow Garden Spider
Argiope aurantia

Conspicuous and large, female black and yellow garden spiders suspended upside down in their webs are familiar sites in gardens, parks, and riparian woodlands in late summer.

IDENTIFICATION The **carapace** is silvery, while the longer-than-wide **abdomen** is boldly marked with black and yellow. The spider's legs are black with variable orange, red, and yellow markings. The much smaller and lighter-bodied male is similarly yet less distinctly marked.

HABITS The female's web consists of a spiral of sticky silk laid on a platform of relatively dry and rigid spokes radiating out from the center. Incorporated into the web is a zigzag ribbon of silk called the **stabilimentum.** Recent studies suggest that it may have reflective qualities that attract flying insect prey to the web.

REPRODUCTION The marble-size and pear-shaped egg sacs are constructed from tough, papery brown silk and suspended in the web. Eggs hatch in early summer. The spiders reach adulthood by August and September and die with the onset of cooler temperatures in early fall.

SIMILAR SPECIES Female banded garden spiders (*A. trifasciata*) are mostly white to pale yellow with narrow bands across the abdomen. Silver garden spiders (*A. argentata*) have abdomens that are nearly as wide as long and are silvery or white in front and flanked on each side by three lobes. Distributed across the southern United States, they construct stabilimenta that form an "X" in the middle of the web.

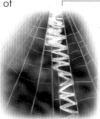

Of Interest Black and yellow garden spiders have been known to occasionally capture and eat small vertebrates like geckos and hummingbirds, in addition to insects.

Stabilimentum

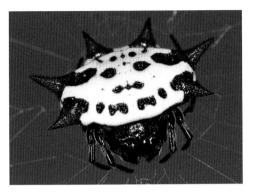

- Size: L male 2.0–3.0 mm; female 5.0–10.0 mm
- Family: **Araneidae**
- Life cycle: **One generation produced annually**
- Range: **Southeastern Virginia to Florida, west to southern California**
- Food: **Flying insects**

Spiny-Backed Orbweaver
Gasteracantha cancriformis

Also known as star or jewel box spiders, these orbweavers spend most of the day hanging head-down in their webs.

IDENTIFICATION Adult females have flat, broad, and brightly colored **abdomens** with six stout reddish or black spine-like projections. Both the **carapace** and legs are dark brown to brownish black. The **dorsal** surface of the hard abdomen is red, white, orange, or yellow, while the underside is black and speckled with yellow. The elevated **spinnerets** found underneath the abdomen are surrounded by a distinct ring. The small, gray, and beetlelike males are seldom seen.

HABITS Found in late summer and fall throughout most of its range and year-round in southern Florida and Texas, this spider is an uncommon resident along coastal southern California. Females hang in orb webs spanning 15 inches or more among shrubs and low tree branches in scrubby woods and along woodland edges. Their bright enameled appearance and sharp projections are thought to deter attacks from birds. The diminutive males are sometimes found suspended on a single strand of silk near the female's web.

REPRODUCTION The flattened, oblong egg sacs are loosely woven with multiple colors of silk threads. They have a single green stripe and are attached to the undersides of nearby leaves. Males die shortly after mating, while females expire soon after producing their eggs.

SIMILAR SPECIES This distinctly spiny arachnid cannot be confused with any other North American spider.

Of Interest During courtship, male spiny-backed orbweavers vibrate their webs in a rhythmic pattern to attract a female for mating.

Alternative color form

- Size: L male 12.0–13.0 mm; female 14.0–16.0 mm
- Family: Oxypodidae
- Life cycle: One generation produced annually
- Range: Virginia to Florida, west to California
- Food: Insects, especially bees, wasps, and caterpillars

Green Lynx Spider
Peucetia viridans

Green lynx spiders are wonderfully adapted for hunting insects among bright-green vegetation.

IDENTIFICATION They are bright green with long pale green or yellow legs that have many long black spines and spots. Reddish spots are found near the eyes and elsewhere on the body. The gradually tapered **abdomen** has three cream chevrons margined in reddish brown that point toward the **cephalothorax.**

HABITS Green lynx spiders inhabit low shrubs and other herbaceous growth in parks, gardens, vacant lots, old fields, thorn scrub, and other habitats and are frequently encountered in agricultural fields, where they are considered natural controls of various pest insects. Relying on speed and their ability to jump, these spiders pounce on bees, wasps, **caterpillars,** and other insects and envelop them with their long, spiny legs.

REPRODUCTION In fall, the smaller, lighter-bodied males seek the larger, heavier-bodied females. Females lay 25 to 600 bright-orange eggs in a yellowish or brownish sac and attach the sac to upper branches of vegetation and shrubs. They remain with the sac until the eggs hatch, tenaciously guarding it, and will spit venom from their fangs to defend their brood. Hatching spiderlings disperse by **ballooning** in spring, reach maturity in summer, and mate in fall.

SIMILAR SPECIES *Peucetia longipalpis* is similar in its appearance and habits to the green lynx spider but has relatively shorter legs and a shorter, more rounded abdomen that lacks chevron markings. This species ranges from southern California to western Texas.

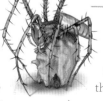

Of Interest Ballooning spiderlings have been collected as high as 16,000 feet in the air and by ships in the middle of the ocean.

Female guarding egg sac

- Size: L male 2.0–4.0 mm; female 5.0–11.0 mm
- Family: **Thomisidae**
- Life cycle: **One generation produced annually**
- Range: **Across southern Canada and United States**
- Food: **Flower-visiting insects**

White-Banded Crab Spider

Misumenoides formosipes

*White-banded crab spiders are sit-and-wait **predators** that hide among flowers growing in gardens, parks, woodland edges, and other open habitats to ambush bees, flies, and other insects.*

IDENTIFICATION Over time, the heavy-bodied females can, within limits, change their overall color to match their backgrounds, and thus may be yellow, yellowish brown, creamy white, or pinkish. The smaller, more slender males cannot change colors. Both sexes have a promi-nent white ridge above the eyes that inspired the common name for this species. The sides of the **car-apace** are distinctly darker, while the **abdomen** is typically marked with black or red, or is some-times plain. Dark markings appear on the **dorsal** surfaces of the enlarged first and sec-ond pairs of legs.

Yellow form

HABITS White-banded crab spiders live up to their name not only because their enlarged legs look somewhat like crab claws but also because they demonstrate an ability to run forward, backward, and sideways with equal rapidity. Females do not build webs or retreats. They hunt for prey among flowers, especially members of the composite family such as golden-rods, asters, and sunflowers.

REPRODUCTION Males spend much of their time searching for the relatively sedentary females. Up to 100 eggs are deposited in white silken sacs attached to vegetation in late summer. Eggs hatch in spring, and the young spiders mature in midsummer.

SIMILAR SPECIES The female golden-rod crab spider (*Misumena vatia*) is similar in overall color, habits, and distribu-tion but lacks the promi-nent ridge over the eyes and distinct markings on the carapace and legs. This female also has the ability to change its overall color to match its background.

- **Size:** L male 6.0–13.0 mm; female 6.0–15.0 mm
- **Family:** Salticidae
- **Life cycle:** One generation produced annually
- **Range:** Southern Canada and United States
- **Food:** Insects

Bold Jumping Spider
Phidippus audax

Large, robust, and brash, bold jumping spiders always attract attention when encountered in and around the home.

IDENTIFICATION The **cephalothorax,** legs, and **abdomen** of mature spiders are typically black, while the **chelicerae** are metallic green; southern populations may have white markings on the **carapace.** The legs may or may not have white bands. Larger females have three white or orange spots on the abdomen and a central white spot or triangle followed by two pairs of white spots on the abdomen, sometimes with a white band across the base. Smaller males have a white band across the base of the abdomen followed by a pair of white spots. Immature spiders may have yellow or orange spots.

HABITS A widespread species most common in the eastern half of the United States, these spiders are found under boards, on tree trunks, and on the sides of buildings. When threatened, they quickly retreat into cracks and crevices. They are sometimes common in agricultural fields, where they are considered important **predators** of stink bugs, beetles, moths, **caterpillars,** and other pest insects. Subadults emerge in spring and reach maturity in summer.

REPRODUCTION After mating, females lay orange eggs in a lens-shaped sac hidden under bark.

SIMILAR SPECIES *Phidippus* contains 60 species and includes the largest and most conspicuous jumping spiders in North America. Many have red or orange abdomens, and nearly all have iridescent chelicerae.

Of Interest Spiderlings of this species build silken winter retreats in curled leaves and old logs. Up to 30 may gather to overwinter in a single shelter.

Adult male with metallic fang bases

- Size: L 9.0–20.0 mm
- Family: Agelenidae
- Life cycle: One generation produced annually
- Range: Across southern Canada and United States
- Food: Insects

Grass Spiders
Agelenopsis species

The horizontal sheet webs of grass spiders that festoon lawns, low bushes, rock walls, and the corners of buildings are especially conspicuous on dewy autumn mornings.

IDENTIFICATION The **carapace** of these spiders is yellowish to brown with a pair of dark stripes, while the **abdomen** is mostly yellowish gray to reddish brown with a variable **dorsal** stripe down the middle. The long hind **spinnerets** are conspicuous and extend well beyond the tip of the abdomen. The legs are faintly marked with narrow rings.

HABITS The grass spiders' horizontal sheet webs are constructed from nonsticky silk, and they include not only a funnellike retreat, where the spider spends most of its time, but also an irregular web above the sheet that entangles flying insects, forcing them onto the sheet. Sensing the vibrations of struggling prey, spiders rush out from their retreats to capture and kill the hapless insect and then haul it back into the retreat. Some species commonly build their webs in and around barns and cellars, and they will often enter homes to establish their webs in dark, undisturbed corners. Adult males spend most of their time searching for females. During the breeding season, males and females are sometimes found together in the same web.

REPRODUCTION The disc-shaped egg sacs of these spiders are produced in fall and hidden in crevices to mature. Females typically die after producing their egg sacs, but individuals of some species may live for several years

SIMILAR SPECIES There are 13 species of *Agelenopsis* in North America, all of which closely resemble one another. Telling them apart is difficult, since they are best distinguished by detailed examination of the male's sexual organs.

Peeking out of silken retreat

- Size: L 4.0–6.0 mm
- Family: Phalangiidae
- Life cycle: One generation produced annually
- Range: Across southern Canada and non-desert regions of United States
- Food: Small arthropods and decaying vegetation

Brown Harvestman
Phalangium opilio

Established throughout the Northern Hemisphere and New Zealand, this is the world's most widespread harvestman.

IDENTIFICATION Brown harvestmen, also known as daddy long-legs, are not spiders and are readily distinguished by having only one distinct body region (spiders have two), a distinctly segmented **abdomen,** and pincherlike **chelicerae.** This arachnid has a broad, dark stripe down the back and long, dark legs with pale bases. Adults of both sexes have long **palps** and spines covering their eye **tubercle,** legs, and bodies. Males have long processes on their palps that project forward, but those of females are not so developed.

HABITS Although found in a wide variety of natural habitats (except deserts), brown harvestmen are most abundant in suburbs, parks, and croplands. They are most often encountered on walls, fences, tree trunks, and other vertical surfaces. On hot days, they sometimes congregate on the shady sides of homes and outbuildings. Harvestmen use their long and slender legs for walking, breathing, smelling, and capturing small prey. Their second pair of legs is used primarily as sensory, rather than ambulatory, organs.

REPRODUCTION Unlike spiders, harvestmen engage in direct copulation. With an extendable egg-laying tube, or **ovipositor,** females will deposit up to several hundred eggs at a time in moist soil, mosses, and rotten wood.

SIMILAR SPECIES Both sexes of *Leiobunum vittatum,* found throughout most of eastern North America, have a dark stripe down the body that becomes less distinctive with age. They have pale legs with dark bands, and the males have less-developed chelicerae.

Of Interest Harvestmen are not venomous and are harmless to people and pets.

Close-up of body showing eyes and segmented abdomen

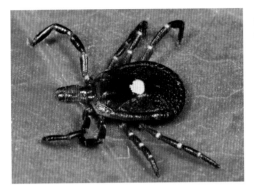

- Size: L to 5.0 mm (unengorged)
- Family: Ixodidae
- Life cycle: One generation produced annually
- Range: New England to Florida, west to Iowa and Texas
- Food: Blood of humans and other mammals, birds

Lone Star Tick
Amblyomma americanum

Lone star ticks are so named because the adult female has a distinct spot on the back that may have a somewhat golden, bronze, or turquoise iridescence.

IDENTIFICATION Adults are rounded and have eyes, long mouthparts, a spot on the **scutum,** and 11 rectangles **(festoons)** along their posterior margin.

HABITS Overwintering eight-legged **nymphs** and adults of lone star ticks emerge in spring to feed, while the six-legged larvae appear in late summer. Lone star ticks are typically encountered in wooded areas with thick underbrush and have achieved pest status with the rapid increase of white-tailed deer populations near homes. This species requires three different hosts to complete its life cycle, one each for larva, nymph, and adult. It bites humans, pets, and other wildlife in the larval,

nymphal, and adult stages. Larvae prefer smaller animals, while nymphs and adults tend to bite larger species. All stages bite humans. Nymphs and adults are important **vectors** of bacterial pathogens that cause human monocytic ehrlichiosis, southern tick-associated rash illness, anaplasmosis, and tularemia—but not Lyme disease.

SIMILAR SPECIES Adult female lone star ticks are distinguished from all other North American ticks by the lone spot, while the males have spots or streaks along the outer margins of the body. Ten species of *Amblyomma* occur in the United States, and all but *A. americanum* are restricted mostly in southern states and do not typically bite humans.

Of Interest Recent research has demonstrated that the bite of this tick in the southeastern United States can cause an allergy to beef, pork, and lamb in humans known as alpha-gal.

Male

DEALING WITH TICKS

The best way to prevent tick-borne diseases is to avoid tick bites altogether. It is good to be vigilant for ticks year-round, especially during the spring and summer months.

AVOID CONTACT AT HOME AND BEYOND

Remove leaf litter, frequently mow the lawn, and clear tall grass and brush around the home and lawn edges. Establish a 10-foot barrier of gravel or wood chips between lawns and wooded areas. Tidy up and secure yards to discourage wild and stray animals, including rodents, from introducing ticks into your yard. Keep decks and playground equipment away from shrubs and woodland edges. Consult with local health and agricultural officials to determine how and when to use pesticides. Whenever outdoors, minimize your contact with ticks by avoiding brushy areas with tall grass and leaf litter and walk along the center of trails.

Tucking pants into tall socks while hiking is one good way to reduce the chances of a tick bite.

When removing a tick, it's important to detach the entire tick. Using tweezers helps ensure that the tick's mouthparts aren't left behind.

TICK REPELLENTS
Whenever working or hiking in tick country, use repellents that contain 20–30 percent DEET on your skin and clothing for protection that lasts several hours. Follow product directions carefully, and avoid getting repellent on your hands or in your eyes and mouth.

PROTECTIVE CLOTHING
Treat clothing (pants and socks), boots, and gear (tents and sleeping bags, for example) with 0.5 percent permethrin, following product instructions carefully. Treated articles will remain repellent to ticks for several washings. Clothing pretreated with permethrin sold through various outlets will repel ticks after many washings.

TICK CHECK
When returning from an outdoor activity and before going indoors, carefully examine your outdoor gear (including coats and packs) for ticks, along with any pets that accompanied you. Once inside, wash all your clothing in hot, soapy water or tumble in a dryer on high heat for an hour. Bathe or shower immediately to wash off any crawling ticks. Then conduct a full-body search in front of a full-length or hand-held mirror to check all nooks and crannies. Inspect behind the knees, between the legs, around the waist, inside the navel, under the arms, around the ears, and on the scalp.

HOW TO REMOVE A TICK
Using fine-tip forceps, grasp the tick close to your skin. Steadily pull the tick upward without twisting or jerking: You want to avoid leaving the tick's mouthparts behind, in the wound. If you can't easily remove the mouthparts with a clean forceps, leave them in your skin and let the wound heal. Wash your hands and the bite site with rubbing alcohol, hot soapy water, or an iodine scrub. Dispose of live ticks in sealed containers with alcohol or by wrapping them in tape to prevent their escape. Avoid crushing a tick with your fingers to avoid infections.

For more information on tick-borne diseases and their prevention, visit *cdc.gov/ticks*.

- **Size:** L to 5.0 mm (unengorged)
- **Family:** Ixodidae
- **Life cycle:** One generation produced every 3 to 24 months
- **Range:** Eastern North America and California
- **Food:** Blood of humans and of wild and domestic mammals

American Dog Tick
Dermacentor variabilis

*The American dog tick is the primary **vector** of Rocky Mountain spotted fever in eastern North America.*

IDENTIFICATION Adults are oval in shape and have eyes, an ornately patterned **scutum,** and 11 rectangles **(festoons)** along their posterior margin. The scutum covers almost the entire body of the male, but only the front third of the female.

HABITS Three different hosts are required for this species to complete its life cycle, which may take 3 to 24 or more months to complete. Adult males and females are active in spring and summer in open habitats with tall grass or low, brushy vegetation. They seek medium- to large-size hosts, such as opossums, raccoons, and coyotes, as well as humans and pets. Males feed briefly but do not become engorged.

REPRODUCTION After mating on the host, the blood-engorged female drops to the ground and produces a mass of up to 4,000 eggs. The six-legged larvae hatch in about a month and begin searching for a suitable small mammal host, especially rodents. The eight-legged **nymphs** are most abundant during the late spring and summer. Since three different hosts are required to complete the life cycle, this (like the lone star, p. 215) is a three-host tick. Adults and nymphs may transmit bacterial pathogens that cause anaplasmosis in cattle, goats, and sheep. A neurotoxin in their saliva causes tick paralysis.

SIMILAR SPECIES The Pacific coast tick *(D. occidentalis)* is most common in California. The Rocky Mountain dog tick *(D. andersoni)* occurs in the mountainous regions of western North America.

Of Interest Rocky Mountain spotted fever is a bacterial infection that is primarily transmitted by a tick bite. Symptoms include severe headache and high fever, and a spotty rash that gives the disease its name. It can be treated with antibiotics.

Centipedes & Millipedes
Subphylum Myriapoda
Centipedes (Class Chilopoda) are somewhat flattened and fleet-footed **predators,** while millipedes (Class Diplopoda) are relatively slow-moving herbivores with bodies that are cylindrical or somewhat flattened in cross section.

- Size: **L to 35.0 mm**
- Family: **Scutigeridae**
- Life cycle: **One generation produced every three years**
- Range: **Across southern Canada and United States**
- Food: **Household insects, spiders, and other small arthropods**

House Centipede
Scutigera coleoptrata

House centipedes are found outdoors on humid summer nights hunting for insects attracted to lights.

IDENTIFICATION Their flattened pale bodies have dark stripes down the back and are borne on 15 pairs of long, banded legs. The large head bears a pair of dark and bulging **compound eyes.**

HABITS House centipedes occur outdoors in moist, cool, protected areas, especially under rocks and logs, and in homes, schools, and office buildings, where they may get trapped in sinks or tubs. Bites inflicted via modified front legs are rare and

result in temporary local pain. They hunt mainly at night for arthropod prey, especially ants, bedbugs, clothes moths, cockroaches, silverfish, termites, and other house insect pests: They should be considered beneficial. Large indoor populations indicate insect infestations. To run rapidly, their respiratory system uses copper-rich proteins **(hemocyanin)** to transport oxygen throughout their bodies. De-oxygenated hemocyanin gives their legs and **antennae** a bluish color.

REPRODUCTION Females live several years and can produce more than 100 offspring in their lifetime.

SIMILAR SPECIES Leggy house centipedes cannot be confused with any other species.

Close-up of head

- Size: **L to 23.0 mm**
- Family: **Paradoxosomatidae**
- Life cycle: **One generation produced annually**
- Range: **North America**
- Food: **Decaying plant tissues**

Greenhouse Millipede
Oxidus gracilis

Greenhouse millipedes, also known as hothouse millipedes, thought to be indigenous to Japan, are now widely established around the world in both tropical and temperate climates.

IDENTIFICATION Their backs are relatively flat, and each body segment has a transverse groove and sides that extend out into lateral keels called **paranota.** The paranota of segments 8 and 9 are blunt posteriorly and do not extend past the posterior margin of the segment. They are deep chestnut brown to black with pale paranota and legs. Adults have 30 (males) or 31 (females) pairs of legs.

HABITS These slow-moving animals are commonly found under rocks and logs in moist gardens, parks, and woodlands.

REPRODUCTION A single female may deposit up to 300 creamy yellow or brown spherical eggs in the soil. Young millipedes hatch with only seven body segments and three pairs of legs. Seven **molts** are required to reach adulthood, and additional segments and pairs of legs are added with each molt. Heavy rains will force these millipedes to climb walls or move inside homes and other structures to seek safety, often in large numbers, where they will eventually die. They do not bite, carry disease, or pose any harm to people or pets, nor do they harm living plants.

SIMILAR SPECIES *Asiomorpha coarcta*, an Asian species established in the Gulf Coast states, is very similar in size and color, but the paranota of body segments 8 and 9 are sharp posteriorly and extend well beyond the segment's posterior margin.

Of Interest When threatened, many millipedes coil their bodies and release a noxious odor.

Adult coiled up

Pillbugs & Sowbugs

Subphylum Crustacea

Also known as roly polys or woodlice, these entirely terrestrial crustaceans have seven pairs of legs. They eat all kinds of plant, fungal, algal, and animal tissues.

- Size: L to 18.0 mm
- Family: Armadillidiidae
- Life cycle: Multiple generations produced annually
- Range: Europe; established across southern Canada and United States
- Food: Plant, animal, fungal, and algal tissues

Common Pillbug

Armadillidium vulgare

Pillbugs are related to lobsters, crabs, crayfish, and shrimp.

IDENTIFICATION They are oblong, slate gray, or black with occasional patches of yellow, and they have seven pairs of legs. The head is not notched in front, and the overlapping plates of the **carapace** are mostly smooth and strongly arched. The posterior margin of the body is rather uniform in outline.

HABITS Common pillbugs occur in moist areas near dwellings and fields and are sometimes considered pests in greenhouses and mushroom farms. They defend themselves by rolling up into a tight ball. Although adapted to life on land, they must maintain proper water balance for respiration and waste elimination. They may live one or two years.

REPRODUCTION After mating, females release up to 100 eggs in a fluid-filled sac, or **marsupium,** under the **abdomen.** Pale juveniles hatch in two or three months and become darker and larger with each **molt.** During molting, pillbugs shed only half the **exoskeleton** at a time.

SIMILAR SPECIES *Armadillidium nasatum* has distinctly patterned carapace plates.

Of Interest Bluish purplish individuals are infected with **iridoviruses,** which attack invertebrates but don't harm people or pets.

Female rolled up with young

- Size: L to 20.0 mm
- Family: **Porcellionidae**
- Life cycle: **Multiple generations produced annually**
- Range: **Europe; established across southern Canada and United States**
- Food: **Plant, animal, fungal, and algal tissues**

Rough Sowbug
Porcellio scaber

Also known as potato bugs, rough sowbugs are omnivorous scavengers that sometimes congregate in large numbers under loose bark.

IDENTIFICATION These small, oblong arthropods are gray or brown in color, occasionally marked with lighter patterning. They have seven pairs of legs. The front margin of the head appears to have three lobes. Females and juveniles tend to be mottled and lighter in color. The overlapping plates of the **carapace** are roughly sculptured and somewhat arched. A pair of appendages extends well beyond the posterior margin of the body.

HABITS Lacking a waxy covering on their **exoskeleton** to help prevent dessication, they typically seek out damp areas in gardens and woodlands under debris and leaf litter on the ground as well as under the loose bark of dead or injured trees.

Mainly **nocturnal** in warmer, hotter weather, they will venture out on cooler, wetter days. Their roughened carapaces minimize the effects of surface tension, and so they are less likely to stick to wet surfaces. Likewise, wet particles are less likely to stick to them.

REPRODUCTION Females mate with many males. Like pillbugs, they brood their eggs in a fluid-filled **marsupium** located on the underside of the **abdomen.** Bluish or purplish individuals are infected with **iridoviruses,** which infect many kinds of invertebrates but are harmless to people and pets.

SIMILAR SPECIES The common name "sowbug" is applied to multiple species in several genera across three different families of isopod crustaceans that cannot roll up into a ball.

Of Interest Unlike the pillbugs they resemble, rough sowbugs are incapable of rolling up into a ball.

Infected with indovirus

Glossary

abdomen Posterior body region in insects and arachnids.

adenosine triphosphate (ATP) Transports chemical energy within cells for metabolism.

adventive Not native; an organism transported to a new habitat.

alate Winged, or an insect during the winged phase of its development.

ametabolous Development with little outward change, other than size.

anamorphosis Type of development where juveniles possess fewer abdominal segments than do adults.

antenna (pl. antennae) Paired sensory appendages located on the head above the mouthparts in insects, crustaceans, millipedes, and centipedes.

antennomere An article, or "segment," of the antenna.

aposematic Possessing distinctive, often contrasting color patterns that serve a defensive purpose by warning predators of unpalatability.

asexual Without sex; often used with reference to reproduction.

ballooning When spiders pay out silk from their spinnerets and wind currents carry them aloft.

Batesian mimicry Phenomenon where palatable species evolve physical features and behaviors similar to those of unpalatable species.

biological control The use of predators, parasites, or pathogens to control pests.

biomass Total weight of organisms living at a particular time or place.

brochosome Chalky substance found on the wings of some leafhoppers.

budding Establishment of a new colony from an existing colony, as in some ants.

carapace Dorsal plate covering the cephalothorax upon which the eyes are located in spiders; or the hard dorsal covering in crustaceans.

carton Paperlike nest material manufactured by some termites and wasps.

caterpillar Soft-bodied larva of butterflies, skippers, and moths.

cephalothorax Body region consisting of head and thorax in arachnids and crustaceans and covered dorsally by the carapace.

cercus (pl. cerci) A paired appendage on the tip of some insect abdomens.

chelicera (pl. chelicerae) The pincherlike mouthparts of arachnids and their relatives; modified into fangs in spiders.

clypeus Area between the front row of eyes and the anterior edge of the carapace in spiders; or the plate above the mouths of insects.

complex As in species complex, a group of closely related and often difficult-to-distinguish species.

compound eye Primary organ of sight in many arthropods consisting of multiple facets or lenses.

cornicle One of a pair of dorsal tubelike structures located toward the tip of the abdomen in many aphids.

crepitate To make a crackling or creaking sound.

crepuscular Pertaining to organisms that are active at dusk or dawn.

cryptic Concealed or hidden.

denticle A small toothlike projection.

detritivore Animals that eat decaying plant, animal, or fungal tissues.

diploid Having chromosomes from both

the male and female parent.

diplosegment A pair of fused body trunk segments in millipedes.

direct development A type of development where the offspring resemble the parents.

diurnal Active during the day.

dorsal Above or on top.

drone A male eusocial bee, especially honey bee or bumble bee.

ectoparasite A parasite that lives outside the body of its host and typically does not kill it.

ectoparasitoid A parasite that lives outside the body of its host and typically kills it.

elytral suture The seam down the back of a beetle where elytra meet.

elytron (pl. elytra) The uniquely modified forewing of beetles.

endoparasite A parasite that lives inside the body of its host and typically does not kill it.

endoparasitoid A parasite that lives inside the body of its host and typically kills it.

envenomation The process of injecting venom via fangs, spines, or stings.

eusocial A social system involving reproductive division of labor (castes) and cooperative rearing of the young by overlapping generations within a colony; includes all ants and termites but only some bees and wasps.

exoskeleton The protective outer covering of arthropods that functions as both skeleton and skin; serves internally as a foundation for muscles and organ systems while externally providing a platform for sensory organs and other structures.

exuvia (pl. exuviae) The cast skin, or

exoskeleton, of an arthropod as a result of molting.

feces Waste or excrement expelled from the anus.

femur (pl. femora) A segment of an insect's leg, third out from where it attaches to the body.

festoons In some ticks, small areas along the posterior abdominal margin distinguished by short grooves.

forb Soft, herblike flowering plant.

gall Abnormal plant growth caused by insects, mites, and other organisms.

gaster The rounded part of the abdomen posterior to the nodelike segment or segments in some ants, bees, and wasps.

halteres In Diptera, the pair of small knobbed structures on the sides of the metathorax, formed from modified hind wings.

haploid Having chromosomes from one parent.

heartwood The dead and usually darker inner layers of a tree trunk in cross section.

hemelytron (pl. hemelytra) The uniquely modified forewing of true bugs.

hemimetaboly Development type also known as gradual or incomplete metamorphosis; stages include egg, nymph or naiad, and adult.

hemocyanin An oxygen-carrying protein found in the hemolymph of crustaceans and select noninsect arthropods.

hemolymph The circulatory fluid of arthropods that is analogous to blood in vertebrates.

holometaboly Development type also known as complete metamorphosis; stages include egg, larva, pupa, and adult.

honeydew A watery and sugary waste product expelled from the anus of aphids, planthoppers, and their relatives.

hypermetamorphosis A type of holometaboly in parasitic insects that develop via successive larval forms differing from one another.

indirect development A type of development where the offspring must pass through a larval stage before reaching adulthood.

instar The stage between molts during nymphal and larval development.

iridovirus A specific group of viruses that infects insects and other arthropods.

jump dispersal The movement of organisms over long distances or across inhospitable habitats.

kleptoparasitic The behavior of an organism that steals prey or food stores of another to feed its own offspring.

labial gland Salivary gland.

luciferase An enzyme present in the light-producing cells of bioluminescent organisms, such as fireflies.

luciferin A compound found in the cells of bioluminescent fireflies and other organisms that, in the presence of an enzyme (e.g., luciferase) and oxygen, produces light.

maggot A legless larva of a fly that lacks a distinctive head.

mandibles The first pair of jawlike mouthparts in insects.

marsupium A fluid-filled pouch in which roly polys brood their eggs.

mesonotum The dorsal plate that covers the middle thoracic segment (mesothorax).

metathorax The last of three thoracic segments that bears the hind legs and hind wings (if present).

microseta (pl. microsetae) A tiny seta, or hairlike external structure, extruding from or surrounding larger primary setae.

molt The act of an insect or other arthropod shedding its exoskeleton in order to grow.

Müllerian mimicry Where multiple species of unpalatable organisms have evolved to resemble one another, their similarities presumably serving to deter predators.

naiad The aquatic, gill-breathing immature stage found in mayflies, dragonflies, damselflies, and stoneflies.

nocturnal Active at night.

node A knoblike or knotlike swelling, often associated with the narrow petiole of ants .

nymph The immature stage of terrestrial insects and aquatic true bugs that develop by hemimetaboly.

ocellus (pl. ocelli) The simple eye of nymphal and many adult insects, usually located on top of the head between the compound eyes.

ootheca (pl. oothecae) A protective egg case formed with secretions produced by female mantises, cockroaches, and grasshoppers.

ovipositor The often visible egg-laying apparatus, especially in female grasshoppers, crickets, and katydids.

palp (pl. palpi, palps) One of a pair of fingerlike appendages of the insect mouth associated with maxilla and labium.

paranota The flangelike sides of the dorsal plates that cover the thoracic, abdominal, or body trunk segments in some insects and other arthropods.

parasite An organism dependent on a host organism for its existence; host is seldom harmed.

parasitoid A parasite that eventually kills its host.

parthenogenetic Able to produce viable eggs and offspring with mating.

peduncle A stalk, petiole, or straplike supporting structure.

petiole A stalk or stem; the narrow stalk or stem by which the abdomen is attached to the thorax in many Hymenoptera; in ants, the nodelike first segment of the abdomen.

pheromones Chemicals produced by special glands and released into the environment to communicate with other members of the same species.

phloem The tissue of vascular plants that transports sugars and other photosynthetic products down from the leaves and throughout the plant.

postmortem interval An estimate of the time of death, which may sometimes be determined by the insect life that has developed on or in a corpse.

predator An animal that preys on other animals.

proboscis The extended beaklike mouthparts of some insects.

prolegs Paired legs present only in caterpillars, attached to abdominal segments forward of the six true legs, which will develop into the mature insect's legs.

pronotal Of the pronotum.

pronotum (pl. pronota) The dorsal plate covering the prothorax.

prothorax The first thoracic segment that bears the front legs; the midsection of the beetle's body.

pruinose Coated with a waxy light-blue, gray, white, or yellowish powder.

pseudergates In primitive termites, nymphs that may or may not have distinctive wing buds and are capable, through further metamorphosis, of developing into adult workers, soldiers, or reproductives.

puncture A tiny pit.

puparium A case formed by the hardening of the last larval skin, in which the pupa is formed in Diptera.

pygidium The last dorsal plate of the abdomen.

queen A female reproductive in an established colony of termites, ants, bees, or wasps.

raptorial Adapted for seizing and holding prey, especially the forelegs of mantises and mantisflies.

reflex bleeding In some beetles, the release of hemolymph laced with defensive chemical compounds through membranes in the leg joints between the femur and tibia.

reproductive Male and female termites, ants, and social bees and wasps that engage in the reproduction activities of a colony.

repugnatorial glands Glands that produce defensive chemical compounds.

saltatorial Adapted for jumping, as in the hind legs of grasshoppers, crickets, and katydids.

sapwood The living and usually lighter outer layers of a tree trunk in cross section.

scale A flattened seta.

scutellum A more or less triangular plate behind the pronotum, especially in beetles and true bugs.

scutum The shieldlike plate in ticks that partially (nymphs, larvae, adult females) or almost completely (adult males) covers the body.

seta (pl. setae) In insects, the hairlike or scalelike structures on the surface of the exoskeleton.

spermatophore A capsule produced by some male insects that contains sperm, nutrients for the female, and sometimes chemical compounds that influence the female's mating and egg-laying behavior.

spinnerets A structure through which silk is extruded through spigotlike openings, sometimes fingerlike in shape.

spiracle An external opening of the tracheal system; a breathing pore.

stabilimentum A dense band of silk usually placed at the center of the web by some orbweaver spiders.

stridulation Sound production by rubbing one body surface against another, usually filelike spines or bumps across a ridge or series of ridges

submarginal The surface just within the margin or edge of a structure.

subphylum (pl. subphyla) A major subdivision of a phylum.

subsocial In insects, a reference to species where the adults only briefly care for their nymphs or larvae.

tarsomere An article of the foot that does not bear claws.

tarsus (pl. tarsi) A foot, including the tarsomere(s) and the claw-bearing pretarsus.

tegmen (pl. tegmina) The thickened forewing possessed by mantises, cockroaches, grasshoppers, crickets, katydids, earwigs, planthoppers, and treehoppers.

teneral The condition of a pale and soft-bodied adult that has just molted.

thoracic Of the thorax.

thorax Middle body region of insects bearing legs and wings, and subdivided into the pro-, meso-, and metathoracic segments.

tibia (pl. tibiae) Located between the femur and tarsus, this is a segment of an insect's leg, fourth out from where it attaches to the body.

tracheole The fine terminal branch of a respiratory tube (trachea).

triungulin The active first instar larva of parasitic insects that develops by hypermetamorphosis.

trophallaxis The exchange of oral or anal secretions among colony members of social insects.

tubercle A small raised bump or knob.

tymbal One of a pair of hardened plates in the sound-producing organ of cicadas.

vector In insects and other arthropods, a species that carries disease-causing pathogens and parasites from one organism to another.

xylem The tissue of vascular plants that conducts water and nutrients up from the roots.

Citizen Science

As you get to know and enjoy insects and spiders, you may choose to share your observations with like-minded naturalists and participate in the study of these amazing creatures. There is still so much to learn about the lives of arthropods—their life cycles, favorite foods, preferred habitats, and behavior, to name just a few things. By participating in citizen science projects, you can help further our understanding of the hundreds of species that surround us.

Citizen science projects are coordinated with entomologists and researchers at universities and natural history museums around the world. Usually, participation in these projects involves collecting and submitting observations and/or specimens to a research database. The great thing about these projects is that you don't have to travel far or invest in expensive equipment to participate. In fact, most of these projects involve collecting data right at home or in nearby vacant lots and parks. Many programs have developed online resources for everyone from teachers and scout leaders to master gardeners and naturalists.

Here you'll find a select list of the dozens of long-term citizen science projects that focus on insects and spiders. New citizen science projects are launched all the time, so check in regularly with your local cooperative extension office, museum, or university, or visit the web pages of the Xerces Society (xerces.org), Your Wild Life (yourwildlife.org), and the Natural History Museum of Los Angeles County (nhm.org).

ANTS & BEES
- Bumble Bee Watch: *bumblebeewatch.org*
- The Great Sunflower Project: *greatsunflower.org*
- School of Ants: *schoolofants.org*

DRAGONFLIES & DAMSELFLIES
- Migratory Dragonfly Partnership: *migratorydragonflypartnership.org*

CICADAS
- Periodical Cicada Citizen Science Project: *robdunnlab.com/projects/urban-buzz*

BEETLES
- The Lost Ladybug Project: *lostladybug.org*
- Backyard Bark Beetles: *backyardbarkbeetles.org*
- Firefly Watch: *legacy.mos.org/fireflywatch*

BUTTERFLIES & MOTHS
- Monarchs: *monarchjointventure.org*
- National Moth Week: *nationalmothweek.org*

ARACHNIDS
- Spiders: *nhm.org/site/activities-programs/citizen-science/spider-survey*

Further Reading

Bradley, Richard A. *Common Spiders of North America*. University of California Press, 2013.

Cranshaw, Whitney. *Garden Insects of North America: The Ultimate Guide to Backyard Bugs*. Princeton University Press, 2006.

Eaton, Eric R., and Kenn Kaufman. *Field Guide to Insects and Spiders and Related Species of North America*. Houghton Mifflin, 2007.

Eiseman, Charley, and Noah Charney. *Tracks and Sign of Insects and Other Invertebrates: A Guide to North American Species*. Stackpole Books, 2010.

Evans, Arthur V. *National Wildlife Federation Field Guide to Insects and Spiders of North America*. Sterling, 2007.

Evans, Arthur V. *Beetles of Eastern North America*. Princeton University Press, 2014.

Hogue, Charles L. *Insects of the Los Angeles Basin*. 3d ed. Natural History Museum of Los Angeles County, 2015.

Marshall, S. A. *Insects: Their Natural History and Diversity; With a Photographic Guide to Insects of Eastern North America*. Firefly Books, 2006.

Murray, Thomas. *Insects of New England and New York*. Kollath+Stensaas, 2012.

Wagner, David. *Caterpillars of Eastern North America*. Princeton University Press, 2005.

WEBSITES
- BugGuide: An outstanding resource for correctly identified insect images, taxonomic information, and other online resources. *bugguide.net*
- iNaturalist: A website where you can upload images to be easily viewed and identified by specialists and knowledgeable naturalists from around the world. *inaturalist.org*
- NatureServe Explorer: An authoritative source for information on more than 70,000 plants, animals, and ecosystems of the United States and Canada; includes in-depth coverage of rare and endangered species. *natureserve.org/explorer*
- U.S. Fish & Wildlife Service Endangered Species: Includes an interactive map with endangered species information by state. *fws.gov/endangered*

ORGANIZATIONS

Entomological Society of America. The largest organization in the world serving the needs of professional entomologists, educators, students, and amateur naturalists. entsoc.org

Xerces Society. An international organization that promotes the conservation of invertebrates. xerces.org

SOURCES FOR BOOKS, EQUIPMENT & SUPPLIES FOR STUDYING INSECTS

Atelier Jean Paquet Inc., 4656, Route Fossamault, Sainte-Catherine-de-la-Jacques-Cartier, Québec G3N 1S8 CANADA; phone 418-875-2276; fax 418-873-1866; atelierjeanpaquet.com

BioQuip Products, Inc., 2321 Gladwick Street, Rancho Dominguez, CA 90220; phone 310-667-8800; fax 310-667-8808; bioquip.com

About the Author

Entomologist Arthur V. Evans, D.Sc., has published 40 scientific papers on the biology of scarab beetles and more than 100 popular articles and books on insects and spiders. His most recent book is *Beetles of Eastern North America*. He was a contributor to *The Book of Beetles*, wrote a chapter in the *National Geographic Illustrated Guide to Wildlife*, and is the author of *National Geographic's Pocket Guide to Insects & Spiders of North America*. His weekly radio segment What's Bugging You? airs Tuesdays on 88.9 FM WCVE Richmond Public Radio. He lives in Richmond, Virginia.

Acknowledgments

I thank Susan Tyler Hitchcock for taking me on board as an author once again, and for the assistance of the National Geographic Books staff, including Sanaa Akkach, Patrick Bagley, Susan Blair, Michelle Cassidy, Judith Klein, and Jennifer Hoff, along with illustrator Mesa Schumacher, photo researcher Kristen Sladen, and the design firm of Grassroots Graphics.

Art Credits

Cathy Keifer/iStockphoto; Spine, Hutton/Tom Stack and Associates/Alamy Stock Photo; 2-3, Carol Polich Photo Workshops/Getty Images; 4, George Grall/Getty Images; 11, Wirepec/iStockphoto; 12 (LE), John Flannery at www .flickr.com/photos/drphotomoto/6141734455, CC license; 12 (RT), Richard Loader/iStockphoto; 13, Kelly/Mooney Photography/Getty Images; 14 (UP), Sari ONeal/SS; 14 (CTR), Syntarsus/SS; 14 (LO), Khlongwangchao/ iStockphoto; 15, Elena Elisseeva/SS; 16 (LE), Carmen Martínez Banús/iStockphoto; 16 (RT), David Tipling/NPL/ Minden Pictures; 17 (UP), Peter Etchells/SS; 17 (LO), alexsvirid/SS; 18 (LE), Rob Routledge, Sault College, Bugwood.org, CC license; 18 (CTR), Whitney Cranshaw, Bugwood.org, CC license; 18 (RT), Bryan Ungard at www .flickr.com/photos/bdungard/14432906418, CC license; 19 (UP LE), Vik Nanda at www.flickr.com/photos/ viknanda/436947700, CC license; 19 (UP CTR), Whitney Cranshaw, Bugwood.org, CC license; 19 (UP RT), Miles Boyer/SS; 19 (LO LE), Joseph Berger, Bugwood.org, CC license; 19 (LO CTR), Kevin Cole at www.flickr.com/ photos/kevcole/3573867281, CC license; 19 (LO RT), tdlucas5000 at www.flickr.com/photos/ tdlucas5000/9353700483, CC license; 21, Joseph Berger, Bugwood.org, CC license; 22, Bob at https://www.flickr .com/photos/mullica/9445421296, CC license; 23 (LE), Marek Velechovsky/SS; 23 (RT), HartmutMorgenthal/SS; 24, Paul Starosta/Getty Images; 25 (UP LE), Whitney Cranshaw, Bugwood.org, CC license; 25 (UP RT), Peggy Greb, USDA Agricultural Research Service, Bugwood .org, CC license; 25 (LO LE), Joseph Berger, Bugwood .org, CC license; 25 (LO RT), Bradley Higbee, Paramount Farming, Bugwood.org, CC license; 27 (UP LE), John Flannery at www.flickr.com/photos/ drphotomoto/4495001988, CC license; 27 (LO LE), John Flannery at www.flickr.com/photos/ drphotomoto/4495001988, CC license; 27 (RT), John Flannery at www.flickr.com/photos/ drphotomoto/4327350635, CC license; 28 (UP), Matt Jeppson/SS; 28 (LO), NoPainNoGain/SS; 29, Photo Researchers/Getty Images; 30, Murat Domkhokov/ iStockphoto; 31 (UP LE), MMCez/SS; 31 (UP CTR), Stephen Bonk/SS; 31 (UP RT), Whitney Cranshaw, Colorado State University, Bugwood.org, CC license; 31 (LO LE), Lorraine Swanson/iStockphoto; 31 (LO CTR), Elliotte Rusty Harold/SS; 31 (LO RT), Steven Katovich, USDA Forest Service, Bugwood.org, CC license; 32 (LE), George Grall/National Geographic Creative; 32 (RT), David Hill at www.flickr.com/photos/dehill/8022217649, CC license; 33 (UP LE), Marek Velechovsky/SS; 33 (UP RT), Joseph Berger, Bugwood.org, CC license; 33 (LO), Robert L. Anderson, USDA Forest Service, Bugwood.org, CC license; 34, Kristian Bell/Getty Images; 35, Top Photo Corporation/SS; 36 (LE), Whitney Cranshaw, Colorado State University, Bugwood.org, CC license; 36 (CTR), Bates Littlehales/National Geographic Creative; 36 (RT), Brian Kunkel, University of Delaware, Bugwood.org, CC

license; 37 (UP LE), Patrick Lamont/SS; 37 (UP RT), Snowleopard1/iStockphoto; 37 (LO LE), Katja Schulz at www.flickr.com/photos/treegrow/8063137530, CC license; 37 (LO RT), Judith Flacke/iStockphoto; 38 (LE), photographer/Getty Images; 38 (UP RT), Lacy L. Hyche, Auburn University, Bugwood.org, CC license; 38 (LO RT), Whitney Cranshaw, Colorado State University, Bugwood .org, CC license; 39 (UP), yogesh_more/iStockphoto; 39 (CTR), Meister Photos/SS; 39 (LO), Jasper Doest/Minden Pictures/National Geographic Creative; 41, Diana Meister/iStockphoto; 42, Arthur V. Evans; 43, Arthur V. Evans; 44, Nature Photographers Ltd/Alamy Stock Photo; 45 (UP), Piotr Naskrecki/Minden Pictures/Getty Images; 45 (LO), Scott Camazine/Science Source/Getty Images; 47, Carol Sharkey/EyeEm/Getty Images; 50 (UP), Dean Pennala/iStockphoto; 51 (UP), Paul Reeves Photography/ SS; 52 (UP), Kyle Horner/SS; 53 (UP), ALAN SCHMIERER at www.flickr.com/photos/sloalan/3883274394, Public Domain; 54 (UP), StevenRussellSmithPhotos/SS; 55, Paul Reeves Photography/SS; 56 (UP), Ian Maton/SS; 57 (UP), Tim Zurowski/All Canada Photos/Getty Images; 58 (UP), Paul Reeves Photography/SS; 59 (UP), Katarina Christenson/SS; 60 (UP), neil hardwick/Alamy Stock Photo; 61 (UP), Nigel Cattlin/Alamy Stock Photo; 62 (UP), Nigel Cattlin/Alamy Stock Photo; 63, HAYKIRDI/ iStockphoto; 64 (UP), Arthur V. Evans; 65 (UP), Arthur V. Evans; 66 (UP), Gallinago_media/SS; 67 (UP), Joseph Berger, Bugwood.org, CC license; 68 (UP), Nature's Images/Science Source/Getty Images; 69 (UP), Grant Heilman Photography/Alamy Stock Photo; 70 (UP), Elliotte Rusty Harold/SS; 71 (UP), Tyler Fox/SS; 72 (UP), Arthur V. Evans; 73 (UP), Joseph Berger, Bugwood.org, CC license; 74 (UP), Nigel Cattlin/Minden Pictures; 75 (UP), Seth Ausubel; 76 (UP), blickwinkel/Alamy Stock Photo; 77, Darlyne A. Murawski/National Geographic Creative; 78, Gerry Bishop/Visuals Unlimited, Inc.; 79 (UP), Patrick Lynch/Alamy Stock Photo; 79 (LO), Gilles San Martin at www.flickr.com/photos/ sanmartin/4815243496, CC license; 80 (UP), john t. fowler/Alamy Stock Photo; 81 (UP), Robyn Waayers; 82 (UP), Mark Brown; 83, Lynette Elliott; 84 (UP), Martin Shields/Science Source/Getty Images; 85 (UP), Jan Miko/ SS; 85 (LO), John W. M. Bush, MIT; 86 (UP), Michael Pettigrew/SS; 87 (UP), Anne Reeves at www.flickr.com/ photos/charlock/15878769435, CC license; 88 (UP), Charles W. Melton; 89 (UP), Arthur V. Evans; 90 (UP), ALAN SCHMIERER at www.flickr.com/photos/ sloalan/4994678730, Public Domain; 91 (UP), Anne Reeves at www.flickr.com/photos/charlock/19669110785, CC license; 92, Gyorgy Csoka, Hungary Forest Research Institute, Bugwood.org, CC license; 93 (UP), Joseph Berger, Bugwood.org, CC license; 94 (UP), Susan Ellis, Bugwood.org, CC license; 95 (UP), Clarence Holmes Wildlife/Alamy Stock Photo; 96 (UP), Russ Ottens, University of Georgia, Bugwood.org, CC license; 97 (UP),

Arthur V. Evans; 98 (UP), Arthur V. Evans; 99 (UP), Alice Abela; 100, Lauree Feldman/Stockbyte/Getty Images; 101, Arthur V. Evans; 102 (UP), Arthur V. Evans; 103 (UP), Arto Hakola/SS; 104 (UP), David Hill at www.flickr.com/photos/dehill/9362006302, CC license; 105 (UP), Nigel Cattlin/Alamy Stock Photo; 106 (UP), David Cappaert; 107 (UP), D. Sikes at www.flickr.com/photos/alaskaent/9233645206, CC license; 108 (UP), Nigel Cattlin/Minden Pictures; 109 (UP), Guillermo Guerao Serra/SS; 110 (UP), alslutsky/SS; 111 (UP), Katja Schulz at www.flickr.com/photos/treegrow/14766045527, CC license; 112 (UP), Tyler Fox/SS; 113, Arthur V. Evans; 114, Arthur V. Evans; 115 (UP), Arthur V. Evans; 116 (UP), Arthur V. Evans; 117 (UP), Arthur V. Evans; 118 (UP), Arthur V. Evans; 119 (UP), Arthur V. Evans; 120 (UP), Arthur V. Evans; 121 (UP), Georgette Douwma/NPL/Minden Pictures; 122 (UP), Arthur V. Evans; 123 (UP), Alice Abela; 124 (UP), Arthur V. Evans; 125 (UP), Arthur V. Evans; 126, Floris van Breugel/Minden Pictures; 127, Gail Shumway, Photographer's Choice/Getty Images; 128 (UP), Gyorgy Csoka, Hungary Forest Research Institute, Bugwood.org, CC license; 129 (UP), Russ Ottens, University of Georgia, Bugwood.org, CC license; 130 (UP), James H. Robinson/Science Source; 131, Todd Ugine; 132 (UP), Arthur V. Evans; 133, Robert and Jean Pollock/Getty Images; 134 (UP), Arthur V. Evans; 135 (UP), Arthur V. Evans; 136 (UP), Stephen Ausmus/USDA-ARS/Public Domain; 136 (LO), Daniel Herms, The Ohio State University, Bugwood.org, CC license; 137 (UP), Maryann Frazier/Getty Images; 137 (LO), Michael Bohne, USDA Forest Service, Bugwood.org, CC license; 137 (UP), Maryann Frazier/Getty Images; 137 (LO), Michael Bohne, USDA Forest Service, Bugwood.org, CC license; 138 (UP), Arthur V. Evans; 139 (UP), Arthur V. Evans; 140 (UP), Bo Zaremba; 141 (UP), Arthur V. Evans; 142, Justin Sullivan/Getty Images; 143, mrfiza/SS; 144 (UP), John Flannery at www.flickr.com/photos/drphotomoto/5180475570, CC license; 145 (UP), hkhtt hj/SS; 146 (UP), Arthur V. Evans; 147 (UP), Joseph Berger, Bugwood.org, CC license; 148 (UP), piyaphat50/iStockphoto; 149 (UP), Arthur V. Evans; 150 (UP), Joseph Berger, Bugwood.org, CC license; 151(UP), Chris Wirth; 152 (UP), John Flannery at www.flickr.com/photos/drphotomoto/5804990428, CC license; 153 (UP), J. T. Chapman/SS; 154 (UP), John Flannery at www.flickr.com/photos/drphotomoto/9364921049, CC license; 155 (UP), Steve Byland/SS; 156 (UP), Arthur V. Evans; 157 (UP), CarolinaBirdman/iStockphoto; 158 (UP), StevenRussellSmithPhotos/SS; 159 (UP), Claudia Steininger/iStockphoto; 160 (UP), Leena Robinson/SS; 161 (UP), John Flannery at www.flickr.com/photos/drphotomoto/5087734770, CC license; 162 (UP), Sari ONeal/SS; 163 (UP), Paul Reeves Photography/SS; 164, Mitch Kezar/Stone/Getty Images; 165, dcwcreations/SS; 166 (UP), John Flannery at www.flickr.com/photos/drphotomoto/10660516284, CC license; 167 (UP), Arthur V. Evans; 168 (UP), Andy Reago & Chrissy McClarren at www.flickr.com/photos/wildreturn/14290943841, Creative Commons license; 169 (UP), Gyorgy Csoka, Hungary Forest Research Institute, Bugwood.org, CC license; 170 (UP), Betty Shelton/SS; 171 (UP), Arthur V. Evans; 172 (UP), StevenRussellSmithPhotos/SS; 173 (UP), Cathy Keifer/iStockphoto; 174 (UP), Matt Jeppson/SS; 175 (UP), Matt Jeppson/SS; 176 (UP), Andy Reago & Chrissy McClarren at www.flickr.com/photos/wildreturn/15122102802, CC license; 177, Scott Linstead/Science Source; 178 (UP), Seth Ausubel; 179 (UP), James Badger at www.flickr.com/photos/jpbadger/3293107343, CC license; 180 (UP), Ian Maton/Shutterstock; 181 (UP), Q Family at www.flickr.com/photos/dasqfamily/3620885694, CC license; 182 (UP), Arthur V. Evans; 183 (UP), Elliotte Rusty Harold/SS; 184 (UP), Elliotte Rusty Harold/SS; 185 (UP), Sharon Moorman; 186 (UP), Arto Hakola/SS; 187 (UP), Elliotte Rusty Harold/SS; 188, FooTToo/iStockphoto; 189, John Flannery at www.flickr.com/photos/drphotomoto/4073255121, CC license; 190 (UP), andipantz/iStockphoto; 191 (UP), Arthur V. Evans; 192 (UP), Rick & Nora Bowers/Alamy Stock Photo; 193 (UP), Barbara Eckstein at www.flickr.com/photos/beckstei/6318861190, CC license; 194 (UP), Robert A. Hamilton, Hamilton Biological, Inc.; 195 (UP), Paul Starosta/Getty Images; 196 (UP), Dave at commons.wikimedia.org/wiki/File:VespulaPensylvanicaWater.jpg, CC license; 197 (UP), George D. Lepp/Getty Images; 198 (UP), Joel Sartore/National Geographic Creative; 199, Nathan Gleave/iStockphoto; 200 (UP), Clemson University - USDA Cooperative Extension Slide Series, Bugwood.org, CC license; 201 (UP), Bryan Reynolds/Visuals Unlimited, Inc.; 202 (UP), Patrick Lynch/Alamy Stock Photo; 203 (UP), Birute Vijeikiene/SS; 204 (UP), Bo Zaremba; 205 (UP), Mark Kostich/iStockphoto; 206 (UP), Rain0975 at www.flickr.com/photos/rain0975/7914467578, CC license; 207, Florida Division of Plant Industry, Florida Department of Agriculture and Consumer Services, Bugwood.org, CC license; 208 (UP), GarysFRP/iStockphoto; 209 (UP), Sam Fraser-Smith at www.flickr.com/photos/samfrasersmith/3778044268, CC license; 210 (UP), David Hill at www.flickr.com/photos/dehill/7868217052, CC license; 211 (UP), John Flannery at www.flickr.com/photos/drphotomoto/3897685791, CC license; 212 (UP), Joseph Berger, Bugwood.org, CC license; 213 (UP), Elliotte Rusty Harold/SS; 214 (UP), Ettore Balocchi at www.flickr.com/photos/29882791@N02/8304379696, CC license; 215 (UP), Arthur V. Evans; 216, RCKeller/iStockphoto; 217, Sebastian Kaulitzki/iStockphoto; 218, Arthur V. Evans; 219 (UP), Joseph Berger, Bugwood.org, CC license; 220 (UP), blickwinkel/Alamy Stock Photo; 221 (UP), Avalon/Picture Nature/Alamy Stock Photo; 222 (UP), D. Kucharski, K. Kucharska/SS.

Index

Boldface indicates illustrations.

A

Achemon sphinxes 178, **178**
American bird
  grasshoppers 68, **68**
American cockroaches 18,
  60, **60**
American dog ticks 218,
  218
American ladies 162, **162**
Antlions 32, 110, **110**
Ants 197–202, **197–202**
  mating swarms 43
  nests 39, **39**
  as social insects 42, 43, 183
  supercolonies 199, **199**
  tending of aphids 197,
    198, 201
  see also Carpenter ants
Apache cicadas 99, **99**
Aphids 106, **106**
  honeydew production
    197, 198, 201
  predators 109, 128, 129,
    131, 145
  reproduction 24
Arachnids 20, 26, 203–218,
  203–218
Argentine ants 197, **197**,
  199
Argus tortoise beetles 135,
  135
Arthropods
  diversity 10, 46
  finding signs of 36–40
  foods 30–34
  habitats 14–15
  how to attract 16–17, 46
  life cycles 24–29
  molting 25
  natural deterrents 46
  observing 16–17
  seasonal activity 12–13
  subphyla 18–23

Asian longhorn beetles
  137, **137**
Asian tiger mosquitoes 36,
  141, **141**, 143
Assassin bugs 32, 36, 86,
  87, **87**, 89

B

Backswimmers 83, **83**
Bagworms 167, **167**
Bedbugs 22, 88, 219
Bees 186–187, **186–187**,
  190, **190**
  colonies 42, 43
  development 28, 28, 183
  how to attract 16, 17,
    188–189
  mouthparts 31, 183
  nests 39, 39, 188
  see also Bumble bees;
    Carpenter bees; Honey
    bees
Beetles 112–139, **112–139**
  detritivores 34
  development 28, 112
  herbivores 31
  how to attract 16
  invasive species 136–137
  mouthparts 31, 112
  nocturnal activity 13, 15
  as pollinators 188
Big dipper fireflies 125,
  125, 127
Biological control 108, 109,
  179, 197
Bioluminescence 125,
  126–127
Bites
  ant 200, 202
  mosquito 142–143
  spider 205, 207
  tick 215, 216, 218

western conenose 88
wheel bug 86
Black and yellow garden
  spiders 37, **41**, 208, **208**
Black-and-yellow mud
  daubers 183, **183**
Black carpenter ants **45**,
  200, **200**
Black saddlebags 52, **52**
Black soldier flies 146, **146**
Black swallowtails **14**, 152,
  154, **154**
  caterpillars **28**, 154, **154**
Black widows 33, **33**, 37, **37**,
  205, **205**
Blister beetles 28, 132, **132**
Blow flies 34, 146, 149
Blue corporal dragonflies
  27
Blue mud daubers 183, **183**
Bluets **55**, 58, **58**
Bold jumping spiders 212,
  212
Boxelder bugs 90, 93, **93**
Broad-headed
  sharpshooters 104, **104**
Broad-winged katydids
  79, **79**
Brown harvestmen 214, **214**
Brown leatherwing beetles
  123, **123**
Brown marmorated stink
  bugs 86, 94, **94**
Brown recluse spiders 207,
  207
Buckeyes, common 161,
  161
Bugs, true 27, 31, 83–108,
  83–108, 151; see also
  Aphids; Assassin bugs;
  Cicadas; Milkweed
  bugs; Stink bugs; Water
  striders; Wheel bugs

Bumble bees 187, **187**
  colonies 44, 187
  declining populations 187
  parasitoid wasp 191
  queens 42, 44, 187
  similar species 190
  subterranean nests 39, 191
  temperature control 187
Butterflies 152–163,
    152–163
  development 28, 152
  how to attract 16, 17, 45,
    165
  mouthparts 31, 152
  Müllerian mimicry 163
  see also Black
    swallowtails; Monarchs;
    Painted ladies; Western
    tiger swallowtails

C

C-9 (nine-spotted lady
    beetle) 131, **131**
Cabbage whites 155, **155**
California prionus 133, **133**
Camel crickets 76
Carnivores 32–33
Carolina grasshoppers
    69, **69**
Carpenter ants 34, **45**, 200,
    200
Carpenter bees 33, 39, 144,
    144, 190, **190**
Carrot beetles 119, **119**
Caterpillar hunter beetles
    32, 112, **112**
Caterpillars. see Bagworms;
    Fall webworms; Gypsy
    moth caterpillars;
    Saddleback caterpillars;
    Tent caterpillars; Tomato
    hornworms; Woolly bears
Cecropia moths 38, **38**,
    175, **175**
Cellar spiders 37, **37**, 203,
    203
Centipedes **22**, 22–23, 32,
    219, **219**

Chinese mantises **33**, 66,
    66
Chokeberry midge
    maggots **31**
Cicada killers 184, **184**, 195
Cicadas 97–101, **97–101**
  annual cicadas 98, **98**, 99,
    99, 101, **101**
  life cycles 29, 100
  periodical cicadas 29, 97,
    97, **100**, 101
  singing 12, 78, 79, 97, 98,
    99, 100
Cockroaches 60–63, **60–63**
  as detritivores 34
  development 27, 60
  eggs and oothecae 36, **36**
  predation by centipedes
    22
  response to light 177
Colorado potato beetles
    138, **138**
  larvae **25**, 138, **138**
Convergent lady beetles
    129, **129**
Cottony cushion scales
    108, **108**
Cow killers 191, **191**
Crab spiders 12, 32, 183,
    211, **211**
Crane flies 140, **140**
Crickets 12, **33**, 68, 73–77,
    73–77, 78
Crustaceans 23; *see also*
    Roly polys

D

Daddy long-legs. see
    Harvestmen
Detritivores 23, 34
Differential grasshoppers
    70, **70**
Doodlebugs 32, 110
Dragonflies and damselflies
    18, 50–59, **50–59**
  development 27, **27**, 50
  mating behavior 55, **55**
  as predators 32

Drone flies 147, **147**, 186
Dung beetles 34, **34**

E

Earwigs 67, **67**, 151
Eastern carpenter bees 33,
    39, 144, 190, **190**
Eastern cicada killers 184,
    184, 195
Eastern hercules beetles
    120, **120**
Eastern lubber
    grasshoppers **18**, 71, **71**
Eastern subterranean
    termites **42**, **43**, 65, **65**
Eastern tent caterpillars
    168, **168**, 169
  cocoons 38, 38, 168
  eggs 36, 36, 168
Ebony jewelwings 55, 59,
    59
Emerald ash borers 136,
    136
European earwigs 67, **67**
European honey bees **31**,
    39, 44, 147, 186, **186**
European hornets **44**, 195,
    195
Eyed click beetles 122,
    122

F

Fall field crickets 73, **73**
Fall webworms 169, **169**
Familiar bluets **55**, 58, **58**
Fiery searchers 112, **112**
Fire ants 42, 201, 202, **202**
Fireflies 125, **125**, 126–127
Five-spotted hawk moths
    176, **176**
Flesh flies 34, 150, **150**
Flies 140, **140**, 144–151,
    144–151
  detritivores 34
  development 28, 140
  how to attract 16
  as pollinators 188

Forensic entomology 34, 146, 150
Four-spotted tree crickets 75, **75**

G

Galls 30, **30**, 31
German cockroaches 62, **62**
Giant eastern crane flies 140, **140**
Golden silk orbweavers 206, **206**
Goldenrod soldier beetles 124, **124**
Grapevine beetles 117, **117**
Grass spiders 213, **213**
Grasshoppers **18**, 68–71, **68–71**
  development 27, 68
  as herbivores 31, **31**
  predation of **33**, 132, 151
  singing 78, 79
Gray hairstreaks 156, **156**
Greater angle-wing katydids 72, **72**
  eggs 36, **36**
Green bottle flies **19**, 149, **149**
Green darners, common 50, **50**
Green fig beetles 121, **121**
Green lacewings **19**, 107, 109, **109**
  larvae **25**, 106, **109**
Green lynx spiders 37, **37**, 210, **210**
Green mantisflies 111, **111**
Green stink bugs **19**, 95, **95**
Greenhouse millipedes **14**, 220, **220**
Greenhouse stone crickets 76, **76**
Greenhouse whiteflies 105, **105**
Gulf fritillaries 157, **157**

Gypsy moth caterpillars **32**, 112, 169

H

Hairstreaks 156, **156**
Harlequin bugs 96, **96**
Harvestmen 203, 214, **214**
Herbivores 30–31
Hercules beetles 120, **120**
Honey bees
  bites and stings 207
  brood cells **24**
  dances 45, **45**
  myth 147
  nests 39
  queens 24, 42
  see also European honey bees
Hornets 40, 42, 44, **44**, 195, **195**
House centipedes **22**, 32, 219, **219**
House flies 146, 148, **148**
House spiders, common 37, 204, **204**

I

Imperial moths 171, **171**
Insect hotels **188**, 189
Insects
  biological control 108, 109, 179, 197
  bioluminescence 125, 126–127
  body regions 18, **20**
  classification 18, 19
  developmental stages 26–29
  eggs 36, **36**
  how to attract 15, 16–17, **17**, 177, **188**, 188–189
  invasive species 96, 136–137, 189, 197
  nests **39**, 39–40
  phototaxis 177
  singing 78–79

social behavior 42–45
Io moths 172, **172**

J

Japanese beetles **19**, 116, **116**
Jerusalem crickets 77, **77**
Jewelwings, ebony 55, 59, **59**
Jumping spiders **21**, 183, 212, **212**
June beetles 114, **114**, 115, **115**

K

Katydids 72, **72**, **79**
  eggs 36, **36**
  singing 12, 72, 78, 79
Kings (caste): role in colonies 42, 43
Kirby's backswimmers 83, **83**
Kissing bugs. see Western conenoses

L

Lacewings. see Green lacewings
Lady beetles **4**, 128–131, **128–131**
  use in pest control 106, 107, 108, 131
  see also Mexican bean beetles; Multicolored Asian lady beetles
Large milkweed bugs **12**, **31**, 89, **89**, 90
Lightning bugs.
  see Fireflies
Little black ants 201, **201**
Lone star ticks 215, **215**
Long-bodied cellar spiders 37, **37**, 203, **203**
Long-legged flies 145, **145**
Long-tailed mealybugs 107, **107**

Lost Ladybug Project (citizen science program) 131
Luna moths 174, **174**

M

Mantises 27, 32, **33,** 36, 60, 66, **66**
Mantisflies 111, **111**
Meadow spittlebugs 103, **103**
Mealybugs 107, **107,** 123
Metamorphosis, types of 26–28
Mexican bean beetles **25, 26,** 130, **130**
Milkweed bugs **12, 31,** 89, **89,** 90, **90**
Millipedes **14,** 22–23, 26, 219, 220, **220**
Mimicry 163, 192
Mites 20, 26, 30, 145
Monarchs **6,** 162, **162**
    caterpillars **29,** 90, **162, 165**
    conservation 165
    how to attract **17,** 165
    migrations 162, 164, 165
    pupae **29,** 90
    threats to 164–165
Mosquitoes 29, 36, **141,** 141–143, **143**
Moths 167–182, **167–182**
    attraction to lights 177, **177**
    cocoons 38, **38**
    development 28, 152
    nocturnal activity 13, 16, 177, 178
    as pollinators 188
Mourning cloaks 158, **158**
Mud daubers 39, 146, 183, **183,** 185, **185**
Multicolored Asian lady beetles 128, **128**
    as introduced species 128, 131
    larvae 25, 128

N

Nature, restoring 46
Nine-spotted lady beetles 131, **131**
Northern dusk-singing cicadas 98, **98**
Northern masked chafers 118, **118**
Northern stick insects 80, **80**

O

Oak treehoppers 102, **102**
Odorous house ants 198, **198**
Oleander aphids 106, **106**
Orbweavers 206, **206,** 209, **209**
    predatory wasps 183, 185
    web construction 32, 206
Oriental cockroaches 61, **61**

P

Pacific dampwood termites 64, **64**
Pacific forktails 57, **57**
Painted ladies **11,** 160
Paper wasps **19, 39,** 40, 42, 44, 194, **194**
Parthenice tiger moths 181, **181**
Pesticide use 46, 94, 189, 216
Phototaxis 177
Pillbugs 23, **23,** 221, **221,** 222
Pipe organ mud daubers 146, 185, **185**
Pipevine swallowtails 152, **152,** 154
Pollinators, how to attract 188–189
Polyphemus moths 38, **38,** 173, **173**
Potato bugs. see Rough sowbugs
Potter wasps 39, 193, **193**

Q

Queen butterflies 162, 163
Queens (caste): role in colonies 42, 43

R

Red admirals 159, **159**
Red imported fire ants 42, 202, **202**
Red milkweed beetles 134, **134**
Regal moths 170, **170**
Repellents
    mosquito 142–143
    tick 217
Roly polys 23, **23,** 34, 221, **221, 222, 222**
Rough sowbugs 222, **222**
Rough woodlice, common **23**

S

Saddleback caterpillars 182, **182**
Scarab beetles 113–121
Seventeen-year cicadas 97, **97**
Sharpshooters 104, **104**
Silky chafers 113, **113**
Silver-spotted skippers **31,** 166, **166**
Silverfish 22, 26, 219
Singing insects 78–79
Skimmers **18,** 51, **51**
Skippers **31,** 166, **166**
Small milkweed bugs 89, 90, **90**
Snowberry clearwings 180, **180**
Southern black widows 33, **33,** 37, 205, **205**
Sowbugs 23, 221, 222, **222**

Spiders 203–213, 203–213
  body regions 20, 20
  egg sacs 37, 37
  spider bites 205, 207
  spiderwebs 32, 33, 40,
    41
  see also Black and yellow
    garden spiders; Black
    widows; Cellar spiders;
    Crab spiders; Green
    lynx spiders; Jumping
    spiders; Orbweavers
Spiny-backed orbweavers
  209, **209**
Spittlebugs 103, 103
Spotted cucumber beetles
  139, 139
Squash beetles 130
Squash bugs 91, 91
Stick insects 80–82, 80–82
Stink bugs
  brown marmorated 86,
    94, 94
  green 19, 95, 95
  harlequin bugs 96, 96
  predators of 86, 212
Striped blister beetles 132,
  132

T

Tachinid flies 151, **151**
Tarantula hawks 192, **192**
Tarantulas 29, 192, **192**
Ten-lined June beetles 115,
  115
Tent caterpillars 112, 168;
  see also Eastern tent
  caterpillars
Termites 64–65, **64–65**
  alates **42**, 64, 65, **65**
  colonies 44, 64, 65
  environmental role 34, 45
  mating swarms 43, 64
  predation by centipedes
    22, 219
  as social insects 42, 43,
    44, 45, 60
  wood damage 64, 65, 200
  workers 43, **43**, 45, **65**

Ticks 215, **215**, 218, **218**
  anamorphosis 26
  bites 215, 216, 218
  how to remove 217, **217**
  protecting against
    216–217
Tiger bee flies 33, 144,
  144
Tomato hornworms **32**, 33,
  176, **176**
Tropical house crickets
  74, **74**
Twelve-spotted skimmers
  18, 51, 51
Two-striped stick insects
  81, **81**

V

Vedalia lady beetles 108
Velvet ants. *see* Cow killers
Viceroys 162, 163, **163**
Vivid dancers 56, **56**

W

Walking sticks. see Stick
  insects
Wandering gliders 53, **53**
Wasps 183–185, **183–185**,
  191–196, **191–196**
  development 28, 183
  gall formation 30, **30**
  how to attract 16, **17**
  nests **39**, 39–40
  social insects 42, 43
  see also European
    hornets; Mud daubers;
    Paper wasps
Water bugs. see Oriental
  cockroaches
Water striders **14**, 84, **84**,
  85, **85**
Western conenoses 88, **88**
Western conifer seed bugs
  92, **92**
Western short-horned stick
  insects 82, **82**
Western tiger swallowtails
  19, 153, **153**

Western yellowjackets 196,
  196
Wheel bugs 32, 36, 86, **86**
White-banded crab spiders
  211, **211**
White-lined sphinxes 179,
  179
Whiteflies 105, **105**, 145
Whitetails, common 51,
  54, 54
Wolf spiders 37
Woodlice **23**, 221
Woolly bears 151, **151**, 181,
  181

Y

Yellow bumble bees 187,
  187
Yellowjackets 40, 42, 44,
  196, **196**

Since 1888, the National Geographic Society has funded more than 12,000 research, exploration, and preservation projects around the world. National Geographic Partners distributes a portion of the funds it receives from your purchase to National Geographic Society to support programs including the conservation of animals and their habitats.

National Geographic Partners
1145 17th Street NW
Washington, DC 20036-4688 USA

Become a member of National Geographic and activate your benefits today at natgeo.com/jointoday.

For information about special discounts for bulk purchases, please contact National Geographic Books Special Sales: specialsales@natgeo.com

For rights or permissions inquiries, please contact National Geographic Books Subsidiary Rights: bookrights@natgeo.com

ISBN: 978-1-4262-1782-1

Printed in China

16/PPS/1

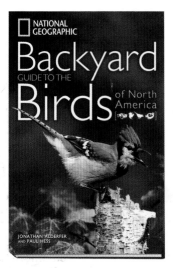